MODFLOW-NWT, A Newton Formulation for MODFLOW-2005

By Richard G. Niswonger, U.S. Geological Survey, Sorab Panday, AMEC
Geomatrix Inc., and Motomu Ibaraki, Ohio State University

Chapter 37 of
Section A, Groundwater
Book 6, Modeling Techniques

Groundwater Resources Program

Techniques and Methods 6–A37

U.S. Department of the Interior
U.S. Geological Survey

U.S. Department of the Interior
KEN SALAZAR, Secretary

U.S. Geological Survey
Marcia K. McNutt, Director

U.S. Geological Survey, Reston, Virginia: 2011

For more information on the USGS—the Federal source for science about the Earth, its natural and living resources, natural hazards, and the environment, visit http://www.usgs.gov or call 1–888–ASK–USGS.

For an overview of USGS information products, including maps, imagery, and publications, visit http://www.usgs.gov/pubprod

To order this and other USGS information products, visit http://store.usgs.gov

Preface

This report describes a Newton formulation of the U.S. Geological Survey's groundwater-flow model (MODFLOW-NWT). MODFLOW-NWT is a standalone version of MODFLOW-2005, including a new Upstream-Weighting (UPW) Package that provides an alternative formulation of the groundwater-flow equation (provided by the BCF, LPF and HUF Packages). MODFLOW-NWT is designed to solve groundwater-flow problems that are nonlinear due to unconfined aquifer conditions, and/or some combination of nonlinear boundary conditions.

All MODFLOW code developed by the U.S. Geological Survey is available for downloading over the Internet from a U.S. Geological Survey software repository. The repository is accessible on the World Wide Web from the U.S. Geological Survey Water Resources Information web page at URL

http://water.usgs.gov/software/lists/groundwater

The performance of MODFLOW-NWT has been tested in a variety of applications. Future applications, however, might reveal errors that were not detected in the test simulations. Users are requested to notify the U.S. Geological Survey of any errors found in this document or the computer program using the email address available on the website mentioned above. Updates might occasionally be made to this document and the MOFLOW-NWT program; users are encouraged to check the website periodically.

Contents

Figures

Tables

Conversion Factors

SI to Inch/Pound

Multiply	By	To obtain
Length		
meter (m)	3.281	foot (ft)
Hydraulic conductivity		
meter per day (m/d)	3.281	foot per day (ft/d)
Hydraulic gradient		
meter per kilometer (m/km)	5.27983	foot per mile (ft/mi)
Transmissivity*		
meter squared per day (m²/d)	10.76	foot squared per day (ft²/d)
Leakance		
meter per day per meter [(m/d)/m]	1	foot per day per foot [(ft/d)/ft]
millimeter per year per meter [(mm/yr)/m]	0.012	inch per year per foot [(in/yr)/ft]

Altitude, as used in this report, refers to distance above the vertical datum.

*Transmissivity: The standard unit for transmissivity is cubic foot per day per square foot times foot of aquifer thickness [(ft³/d)/ft²]ft. In this report, the mathematically reduced form, foot squared per day (ft²/d), is used for convenience.

MODFLOW-NWT, A Newton Formulation for MODFLOW-2005

By Richard G. Niswonger[1], Sorab Panday[2], and Motomu Ibaraki[3]

Abstract

This report documents a Newton formulation of MODFLOW-2005, called MODFLOW-NWT. MODFLOW-NWT is a standalone program that is intended for solving problems involving drying and rewetting nonlinearities of the unconfined groundwater-flow equation. MODFLOW-NWT must be used with the Upstream-Weighting (UPW) Package for calculating intercell conductances in a different manner than is done in the Block-Centered Flow (BCF), Layer Property Flow (LPF), or Hydrogeologic-Unit Flow (HUF; Anderman and Hill, 2000) Packages.

The UPW Package treats nonlinearities of cell drying and rewetting by use of a continuous function of groundwater head, rather than the discrete approach of drying and rewetting that is used by the BCF, LPF, and HUF Packages. This further enables application of the Newton formulation for unconfined groundwater-flow problems because conductance derivatives required by the Newton method are smooth over the full range of head for a model cell.

The NWT linearization approach generates an asymmetric matrix, which is different from the standard MODFLOW formulation that generates a symmetric matrix. Because all linear solvers presently available for use with MODFLOW-2005 solve only symmetric matrices, MODFLOW-NWT includes two previously developed asymmetric matrix-solver options. The matrix-solver options include a generalized-minimum-residual (GMRES) Solver and an Orthomin / stabilized conjugate-gradient (CGSTAB) Solver. The GMRES Solver is documented in a previously published report, such that only a brief description and input instructions are provided in this report. However, the CGSTAB Solver (called χMD) is documented in this report.

Flow-property input for the UPW Package is designed based on the LPF Package and material-property input is identical to that for the LPF Package except that the rewetting and vertical-conductance correction options of the LPF Package are not available with the UPW Package. Input files constructed for the LPF Package can be used with slight modification as input for the UPW Package. This report presents the theory and methods used by MODFLOW-NWT, including the UPW Package. Additionally, this report provides comparisons of the new methodology to analytical solutions of groundwater flow and to standard MODFLOW-2005 results by use of an unconfined aquifer MODFLOW example problem. The standard MODFLOW-2005 simulation uses the LPF Package with the wet/dry option active. A new example problem also is presented to demonstrate MODFLOW-NWT's ability to provide a solution for a difficult unconfined groundwater-flow problem.

Introduction

The finite-difference model MODFLOW-2005 solves the groundwater-flow equation using linear and nonlinear numerical-solution methods (Harbaugh, 2005). Although there are several options available within MODFLOW-2005 to formulate the cell-by-cell flow terms and to solve the resulting linear system of equations, the Picard method is the only method available to solve the nonlinear equations that arise when representing unconfined aquifers and nonlinear boundary conditions. The Picard method is the repeated approximation of a solution to a nonlinear equation wherein each new iteration provides a more accurate solution than the previous iteration. For the case of the groundwater-flow equation, an approximate solution of the groundwater heads calculated for one iteration is used in a subsequent iteration to calculate a more accurate groundwater head solution. Iterations are repeated until the change in groundwater heads is below a user-specified tolerance.

The Newton method is another widely used method for solving systems of nonlinear equations that has been shown to be a useful alternative to the Picard method for many problems (HydroGeoLogic, 1996; Painter and others, 2008; Keating and Zyvoloski, 2009). The Newton formulation described in this report extends the applicability of MODFLOW, especially to those problems representing unconfined aquifers and surface-water/groundwater interaction. The Newton method can be applied to any smooth and continuous function. However, because the linear system of equations for cell-by-cell flow used by the BCF, LPF, and HUF Packages is not continuous

---

[1] U.S. Geological Survey.

[2] AMEC Geomatrix Inc, Herndon, Virginia.

[3] School of Earth Sciences, Ohio State University, Columbus, Ohio.

during drying and rewetting of cells in an unconfined simulation, an upstream-weighting function is developed to smooth cell connections in the discretized groundwater-flow equation, as presented by Painter and others (2008), and Keating and Zyvoloski (2009). Upstream weighting means that if flow is from cell i to cell j in a finite-difference grid, then hydraulic head in cell i alone is used to calculate the horizontal intercell conductance between cells i and j. The upstream-weighting approach avoids groundwater flow out of dry cells, which is not physically realistic and can cause model convergence failure. Therefore, MODFLOW-NWT includes an upstream-weighting (UPW) intercell conductance package as a replacement internal flow package to those provided by the BCF, LPF, and HUF Packages. The UPW Package mirrors the LPF Package other than it uses upstream weighting and only supports a single formulation for calculating vertical conductance. Vertical conductance is calculated as the conductance of two one-half cells in a series with continuous saturation between them (Harbaugh, 2005, p. 5-8). Additionally, optional interpretations of the groundwater storage input variable are not supported by the UPW Package. Consequently, the LPF Package input variable "OPTIONS" is not supported by the UPW Package. Options for averaging intercell hydraulic conductivity and input values and formats are nearly identical among the UPW and LPF Packages. Except for a few minor input changes, input generated for the LPF Package can be used as input for the UPW Package.

Another significant difference between the UPW and LPF Packages is caused by differences between the MODFLOW-NWT and MODFLOW-2005 formulations. A model cell is not set to no flow if it has constant transmissivity for both the MODFLOW-2005 and MODFLOW-NWT formulations. However, MODFLOW-2005 will set a dewatered cell to a no-flow condition if the cell has time-variable transmissivity and the cell is dewatered (Harbaugh, 2005, p. 5–6). Accordingly, cells can be reset to active using the rewetting option in the LPF Package. However, MODFLOW-NWT will not set a dewatered cell to no flow and there is no need for input variables related to rewetting in the UPW Package input file. The UPW Package input file may not contain rewetting data.

The Newton method is a commonly used method in the earth sciences to solve nonlinear equations, such as for solving the multiphase-flow and variably-saturated flow equations (Huyakorn and others, 1986; Pruess, and others, 1999; Panday and Huyakorn, 2004; Maxwell and Miller, 2005). Because many recently developed packages for MODFLOW-2005 apply nonlinear boundary conditions to the groundwater-flow equation (Merritt and Konikow, 2000; Halford and Hanson,

2002; Niswonger and Prudic, 2005; Niswonger and others, 2006; Konikow and others, 2009), the Newton method may improve convergence and computational efficiency when using these packages. Additionally, recent studies indicate that the Newton method is better than the Picard method for solving problems representing unconfined aquifers in which the water table rises and declines through the interface between model layers (HydroGeoLogic, 1996; Painter and others, 2008; Keating and Zyvoloski, 2009). Drying and subsequent wetting of cells can cause convergence failure of the groundwater-flow equation when using the Picard method with the rewetting algorithms of the BCF, LPF, and HUF Packages (McDonald and others, 1991; Doherty, 2001).

A fundamental difference between the Newton and Picard methods is the matrix of equations that is solved during the iterative-solution procedure. The Newton method requires a matrix of partial derivatives to the finite-difference approximations of the groundwater-flow equation; this matrix is called the Jacobian (Patel, 1994). The Picard method requires calculation of a matrix of coefficients that result from the finite-difference approximations to the groundwater-flow equation, called the conductance matrix (Harbaugh, 2005).

Unlike the Picard method, which requires the solution of a symmetric matrix of linear equations when applied to the groundwater-flow equation, the Newton method in MODFLOW-NWT requires solution of an asymmetric matrix. Consequently, the linear solvers presently available for MODFLOW-2005 cannot be used with MODFLOW-NWT. To address this limitation, two iterative, linear-solver options that handle asymmetric matrices are available for use with the Newton Solver Package. The first is based on a preconditioned generalized minimum residual method called GMRES (Barrett and others, 1994; Kelley, 1995; Greenbaum, 1997; and Saad, 2003). The GMRES solver was adapted for HYDROTHERM by Kipp and others (2008) using software originally developed by Saad (1990) and algorithms described in Saad (2003). For more details of the GMRES solver, readers can refer to Kipp and others (2008).

The second option is the χMD solver developed by Motomu Ibaraki, the third author of this report. This solver is based on a preconditioned conjugate gradient type matrix solver including Orthomin and CGSTAB acceleration schemes (van der Vorst, 1992). The preconditioning scheme permits various levels of incomplete LU factorization and reordering of unknowns. Additionally, χMD allows the use of a drop tolerance scheme. The χMD matrix solver option is documented in Appendix C.

Purpose and Scope

This report describes a new computer program (MODFLOW-NWT) for solution of the three-dimensional groundwater-flow equation. MODFLOW-NWT uses the Newton solution method and unstructured, asymmetric matrix solvers to calculate groundwater head, often referred to as a Newton-Krylov method (Knoll and Keyes, 2004). MODFLOW-NWT was designed to work with the UPW Package to solve complex unconfined groundwater-flow simulations including those characterized by drying and rewetting of cells. The UPW Package computes the horizontal-conductance terms for the unconfined groundwater-flow equation in a different manner than do the BCF, LPF, and HUF internal flow packages of MODFLOW-2005, and is a replacement to these three packages for calculating conductance and storage terms in MODFLOW-NWT. Accordingly, MODFLOW-NWT requires two new input files that are not used by MODFLOW-2005. These new input files are (1) the UPW Package input file that contains input required for the internal-flow calculations, and (2) the NWT input file that contains input values required by the Newton and matrix solver methods.

Description of MODFLOW-NWT

There are important functional aspects of MODFLOW-NWT that differ from MODFLOW-2005, including (1) the use of upstream weighting for calculating horizontal conductance of unconfined aquifers, (2) all variable-head cells that are active at the start of a simulation remain active throughout the simulation, (3) the horizontal conductance was modified for unconfined conditions to smooth discontinuities during cell drying and rewetting, (4) the storage formulation was modified to smooth storage changes during cell wetting/drying and during transitions between confined and unconfined conditions, and (5) the storage formulation was modified for unconfined conditions such that there are no changes in storage for head changes that occur beneath the cell bottom.

Most packages supported by MODFLOW-2005 can be used with MODFLOW-NWT without modification, provided that sinks, such as wells, drains, and rivers, do not remove water from dry cells. For example, if the WEL Package is used to remove water from a dry cell, then convergence failure may occur. To alleviate this problem, modifications were made to the WEL Package to reduce the pumping rate to zero in dry cells, as explained in the section Additional Modflow Packages Modified for MODFLOW-NWT. The SFR and UZF Packages also were modified to avoid removing water from dry cells and to calculate derivatives for the Newton

formulation. Although most standard MODFLOW-2005 Packages can be used in MODFLOW-NWT (table 1), poorly conceptualized input conditions for some packages can result in water flowing out of dry cells, possibly resulting in convergence failure. An example of this problem is illustrated by the EVT Package when the extinction depth has a value below the altitude of the cell bottom. This also could cause a problem for MODFLOW-2005; however, in some cases, it can be alleviated when dry cells are made inactive. Dry cells remain active in MODFLOW-NWT, and thus, applying sinks to dry cells is more often detrimental to achieving a solution. MODFLOW-NWT also supports standard MODFLOW-2005 solvers and internal-flow packages using the Picard method (table 1). In this case, MODFLOW-NWT runs identically to MODFLOW-2005.

Table 1. Functionality of version 1 of MODFLOW-NWT.

[All packages listed in this table are supported by MODFLOW-NWT. Some packages were modified for MODFLOW-NWT to improve robustness or to provide cell hydraulic properties from the UPW Package. MODFLOW-NWT also supports standard MODFLOW-2005 solvers and flow packages using the original MODFLOW-2005 Picard method]

	Package abbreviation	Modified for MODFLOW-NWT
Packages used with Newton Formulation		
BAS	Basic	No
CHD	Time-Variant Specified-Head Option	No
DRN	Drain	No
EVT	Evapotranspiration	No
GAG	Gage	No
GHB	General-Head Boundary	No
HFB	Horizontal Flow Barrier	Yes
LAK	Lake	Yes
MNW1	Version 1 Multi-Node Well	Yes
MNW2	Version 2 Multi-Node Well	Yes
NWT	Newton Solver	Yes (new)
OBS	Observation Process	No
RCH	Recharge	No
RIV	River	No
SFR	Streamflow Routing	Yes
UPW	Upstream Weighting	Yes (new)
UZF	Unsaturated-Zone Flow	Yes
WEL	Well	Yes
Packages used with MODFLOW-2005 Formulation		
BCF	Block-Centered Flow	No
DE4	Direct Solver	No
HUF	Hydrogeologic-Unit Flow	No
LPF	Layer-Property Flow	No
PCG	Preconditioned Conjugate Gradient	No
SIP	Strongly Implicit Procedure	No

For unconfined conditions, the terms in the discretized groundwater-flow equation are calculated using upstream weighting of the saturated thickness, rather than a weighted average of the saturated thickness between adjacent cells, as is done in MODFLOW-2005. Rather than solely adding the coefficients of the discretized groundwater-flow equation into the solution matrix and applying the Picard method, derivatives of the groundwater-flow equation also are added to the solution matrix and the resulting nonlinear equations are solved using the Newton method. Application of the Newton method requires several changes to the internal structure of MODFLOW-2005 and its calculations. The following section describes these changes, beginning with the modified calculations of the discretized groundwater-flow equation made by the UPW Package, and followed by the application of the Newton method and corresponding numerical procedures.

UPW Package

MODFLOW-NWT must be used with the upstream weighting (UPW) Package. The UPW Package is an alternative to the BCF, LPF, and HUF Packages for calculating all terms in the discretized groundwater-flow equation. The upstream-weighting approach differs from the approaches used in the BCF, LPF, and HUF Packages in which heads in two adjacent cells are used to calculate the intercell horizontal conductance. Additionally, the UPW Package smoothes the horizontal-conductance function and the storage-change function during wetting and drying of a cell to provide continuous derivatives for solution by the Newton method. Smoothing is applied to both the horizontal-conductance functions and the storage functions using the same curve (fig. 1). The BCF, LPF, and HUF Packages use a linear function to calculate horizontal conductance and storage change (fig. 1). The difference in the two curves shown in figure 1 illustrates the error that is created by the smoothing that is used by MODFLOW-NWT. However, the smoothing interval (Ω) can be made very small (for example, 1×10^{-5} m), such that this error is small.

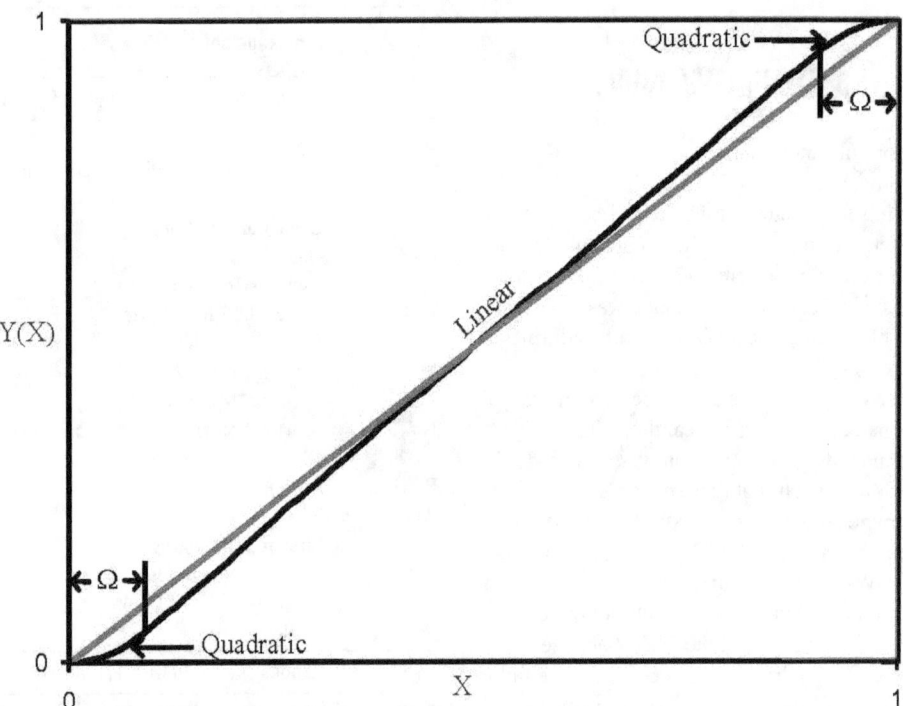

Figure 1.   Combined quadratic and linear functions used to smooth conductance and storage in MODFLOW-NWT (black), and a linear function that is used by MODFLOW-2005 (blue). X is the saturated thickness divided by the cell thickness, Y is the value of the smoothing function, and Ω is the interval of X where the quadratic equation is applied, and is equal to 0.1 in this example (NWT input file variable THICKFACT).

Conductance Equations

An internal flow package of MODFLOW calculates the CV, CR, and CC conductance coefficients and groundwater-storage terms in the finite-difference flow equation written as (equation 2-26 of the MODFLOW-2005 document)

$$
\begin{aligned}
CV_{i,j,k-1/2}h_{i,j,k-1} &+ CR_{i,j-1/2,k}h_{i,j-1,k} + CC_{i-1/2,j,k}h_{i-1,j,k} \\
&+ (-CV_{i,j,k-1/2} - CR_{i,j-1/2,k} - CC_{i-1/2,j,k} \\
&- CV_{i,j,k+1/2} - CR_{i,j+1/2,k} - CC_{i+1/2,j,k} \\
&+ HCOF_{i,j,k})h_{i,j,k} + CV_{i,j,k+1/2}h_{i,j,k+1} \\
&+ CR_{i,j+1/2,k}h_{i,j+1,k} + CC_{i+1/2,j,k}h_{i+1,j,k} \\
&= RHS_{i,j,k},
\end{aligned}
\tag{1}
$$

where i,j,k are indices for the column, row, and layer directions, respectively; $CV_{i,j,k-1/2}$ is the intercell conductance between cells $i,j,k-1$ and i,j,k; $CR_{i,j-1/2,k}$ is the intercell conductance between cells $i,j-1,k$ and i,j,k; $CC_{i-1/2,j,k}$ is the intercell conductance between cells $i-1,j,k$ and i,j,k; $CV_{i,j,k+1/2}$ is the intercell conductance between cells i,j,k and $i,j,k+1$; $CR_{i,j+1/2,k}$ is the intercell conductance between cells i,j,k and $i,j+1,k$; and $CC_{i+1/2,j,k}$ is the intercell conductance between cells i,j,k and $i+1,j,k$. All coefficients of $h_{i,j,k}$ that do not include conductance between nodes (for example, coefficients for head-dependent boundary conditions) are combined into a single term, $HCOF_{i,j,k}$, and all right-hand-side terms are combined into the term $RHS_{i,j,k}$, including storage terms and boundary conditions (Harbaugh, 2005, p. 2-12 and 2-13).

As shown in equation 1, the finite-difference equations that are solved by MODFLOW are the averages of the half-cell conductances between nodes of adjacent cells—that is, "branch conductances"—rather than conductances defined within individual cells. The horizontal-conductance terms CR and CC of equation 1 are calculated between adjacent horizontal nodes. CR terms are oriented along rows and thus specify conductance between two nodes in the same row. Similarly, CC terms specify conductance between two nodes in the same column, and CV terms specify conductance between two nodes in adjacent layers for the same row and column.

An internal-flow package reads data defining the horizontal hydraulic conductivity for individual cells and calculates conductances between nodes. Each conductance is a combination of several parameters used in Darcy's law. Darcy's law defines one-dimensional flow in a prism of porous material as

$$
Q_{1,2} = \left[\frac{KA}{L}\right]_{1,2}[h_1 - h_2],
\tag{2}
$$

where $Q_{1,2}$ is the volumetric-flow rate between nodes 1 and 2, K is the hydraulic conductivity of the material in the direction of flow, A is the cross-sectional area perpendicular to the flow, $h_1 - h_2$ is the head difference across the prism parallel to flow, and L is the length of the prism parallel to the flow path. The conductance $C_{1,2}$ is then defined as (Harbaugh, 2005, eq. 5-4)

$$
C_{1,2} = \left[\frac{KA}{L}\right]_{1,2} = \left[\frac{Kbw}{L}\right]_{1,2},
\tag{3}
$$

where the area A is replaced in the last equality of equation (3) by w, the width of the cell interface, and b, the saturated thickness of the cell interface.

The BCF and LPF Packages provide four methods by which the intercell horizontal-conductance term is calculated, as discussed in Chapter 5. The UPW Package provides an alternative approach for calculating horizontal conductance for a cell with time-variable transmissivity; however, the options for calculating intercell conductance for a cell with constant transmissivity are the same as for the LPF and UPW Packages. The intercell-conductance term for a cell with time-variable transmissivity is calculated from equation (3) by the UPW Package using an average hydraulic conductivity (K_{ave}) multiplied by the upstream saturated thickness (b_{up}), as shown by Keating and Zyvoloski (2009) and Painter and others (2008). This is different from the schemes of the BCF, LPF, or HUF Packages in which the intercell transmissivity, as defined by Kb, is used to calculate the horizontal conductance in equation (3). Accordingly, the UPW Package calculates the average hydraulic conductivity between cells during model initialization, and then averages conductance between cells using the upstream saturated thickness during iteration of the solution scheme as

$$
\frac{b_{up}K_{ave}w}{L}.
\tag{4}
$$

The horizontal row conductance between cells $i,j-1,k$ and i,j,k when upstream weighting is applied is calculated as

$$
CR_{i,j-1/2,k} = \Delta C_i \frac{K_{ave}}{\Delta R_{j-1}}\left[h_{up} - BOT_{up}\right],
\tag{5}
$$

where b_{up} has been replaced by $[h_{up} - BOT_{up}]$; h_{up} is the maximum head of either $h_{i,j,k}$ and $h_{i,j-1,k}$; ΔR_{j-1} is the distance between the center of cells $i,j-1,k$ and i,j,k; ΔC_i is the column width for cell i,j,k; and BOT_{up} is the cell bottom altitude corresponding to h_{up}. Additionally, if h_{up} is greater than TOP_{up} (the cell top altitude corresponding to h_{up}), then the horizontal row conductance is calculated for confined conditions as

$$
CR_{i,j-1/2,k} = \Delta C_i \frac{K_{ave}}{\Delta R_{j-1}}\left[TOP_{up} - BOT_{up}\right].
\tag{6}
$$

Finally, if h_{up} is less than BOT_{up} then the horizontal row conductance is calculated as

$$CR_{i,j-1/2,k} = 0 . \qquad (7)$$

Following Goode and Appel (1992), the intercell hydraulic conductivity (K_{ave}) can be calculated using either a logarithmic, weighted-harmonic, or arithmetic average. For example, a weighted-harmonic average is calculated according to McDonald and Harbaugh (1988) and Harbaugh (2005) as

$$K_{ave} = K_{i,j-1/2,k}$$
$$= 2.0\Delta R_{j-1} \frac{K_{i,j-1,k} K_{i,j,k}}{\Delta R_j K_{i,j-1,k} + \Delta R_{j-1} K_{i,j,k}} , \qquad (8)$$

where $K_{i,j-1,k}$ and $K_{i,j,k}$ are the hydraulic conductivity values for cells $i,j-1,k$ and i,j,k, respectively. For confined conditions, intercell conductance is calculated according to the uniform transmissivity options provided by the LPF Package (Harbaugh, 2005, p. 5-4 to 5-6).

To smooth transitions between equations (5), (6), and (7) as the water level moves upward (or downward) through a cell, the equations are further modified by using a quadratic function over small intervals at the fully-dry and completely-saturated ends of the conductance function (fig. 1). The smoothing is done over the small distance Ω that typically is set to a value less than 1.0×10^{-5} m to avoid errors associated with deviating from the correct (linear) function used by MODFLOW-2005. The smoothed conductance function thus obtained is expressed as

$$
\begin{aligned}
CR_{i,j-1/2,k} &= 1 \times 10^{-9} & & X \leq 0 \\
CR_{i,j-1/2,k} &= \Delta C_i \Delta Z_k \frac{K_{i,j-1/2,k}}{\Delta R_{j-1}} \left[\frac{0.5AX^2}{\Omega} \right], & & 0 < X \leq \Omega \\
CR_{i,j-1/2,k} &= \Delta C_i \Delta Z_k \frac{K_{i,j-1/2,k}}{\Delta R_{j-1}} \left[AX + 0.5(1-A) \right], & & \Omega < X \leq (1-\Omega) \\
CR_{i,j-1/2,k} &= \Delta C_i \Delta Z_k \frac{K_{i,j-1/2,k}}{\Delta R_{j-1}} \left[1 - \frac{0.5A(1-X)^2}{\Omega} \right] & & (1-\Omega) < X < 1 \\
CR_{i,j-1/2,k} &= \Delta C_i \Delta Z_k \frac{K_{i,j-1/2,k}}{\Delta R_{j-1}} & & 1 \leq X , \qquad (9)
\end{aligned}
$$

where $CR_{i,j-1/2,k}$ is the smoothed intercell conductance, $\Delta Z_k = TOP_{up} - BOT_{up}$ is the cell thickness, and X is calculated as $X = \dfrac{(h_{up} - BOT_{up})}{(TOP_{up} - BOT_{up})}$, and $A = \dfrac{1}{1-\Omega}$. Ω is necessary to prevent discontinuities in the value of $CR_{i,j-1/2,k}$; such discontinuities would cause the Newton solution method to fail. Doherty (2001) applied smoothing to the aquifer transmissivity as a cell dewatered using an exponential function. Figure 1 shows the values provided by the bracketed terms in equation 9 over the full range of X.

The horizontal conductances along columns, CC, may be derived in a similar fashion as described above for horizontal conductance along rows, CR. The vertical conductance along layers, CV, is calculated as discussed in Harbaugh (2005, chapter 5).

Treatment of Dry Cells

Another important difference between the UPW Package and the BCF, LPF, or HUF Packages is that dry cells are not set to a no-flow condition in the UPW Package, as they are in the BCF, LPF, and HUF Packages. MODFLOW-NWT will calculate a head in a dry cell that may be greater than an adjacent wet cell, as shown in figure 2. For these circumstances, if arithmetic averaging is used then water will flow out of a dry cell to an adjacent partially saturated cell, which is inconsistent with flow continuity and can cause model-convergence failure. This condition is illustrated in figure 2, which shows flow from a dry cell with a head just slightly greater than 100 ft (that is, a head equal to the bottom of the cell) to a wet cell with a head of 95 ft. To avoid flow out of a dry cell, the UPW Package uses upstream weighting, and, according to equations 5 and 7, the conductance between a dry cell and adjoining wet cell is zero. Thus, the UPW Package can keep a dry cell active while not allowing water to flow out of a dry cell. Similar problems arise during rewetting of cells when harmonic averaging is used. Upstream weighting provides a continuous solution for all unconfined groundwater flow conditions.

Inflow to a dry cell, either from adjacent cells, overlying cells, or an external source simulated by one of the stress packages, automatically flows downward to an underlying cell if there are deeper layers (fig. 3). MODFLOW-NWT makes the assumption of vertical equilibrium between liquid water and air, in which all water occupies the lower portion of the cell, and is assumed to be saturated groundwater storage, and all air (not including water vapor) occupies the upper portion of the cell. Of

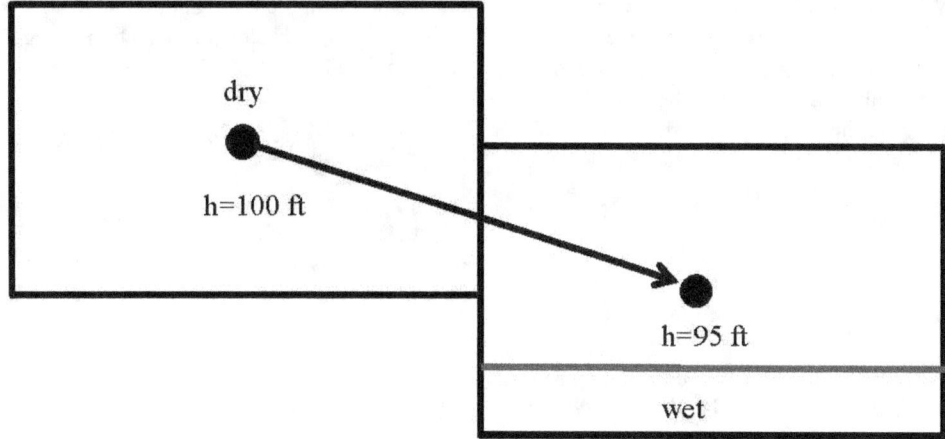

Figure 2. Water flowing out of an active, yet dry cell. MODFLOW-NWT eliminates this flow by setting the conductance between the two cells equal to zero.

course, air is assumed to be passive and all water entering the top of the cell is added to saturated storage, instantaneously. These assumptions make the vertical conductance a constant, regardless of the amount of water stored in a cell. Following the groundwater-flow equation, if a cell is dry (that is, head is below cell bottom) and underlain by a fully or partially saturated cell, horizontal conductance will be zero, and the head in the dry cell can be calculated from the flow into the dry cell in the following manner:

$$Q_{i,j,k+1/2} = Q_{i,j,k}^{in}, \tag{10a}$$

$$Q_{i,j,k+1/2} = CV_{i,j,k+1/2}(h_{i,j,k+1} - h_{i,j,k}), \tag{10b}$$

$$h_{i,j,k} = \frac{Q_{i,j,k}^{in}}{CV_{i,j,k+1/2}} + h_{i,j,k+1}, \tag{10c}$$

where $Q_{i,j,k}^{in}$ is the sum of inflow to cell i,j,k from adjacent cells or from an external source, and $CV_{i,j,k+1/2}$ is the conductance between nodes i,j,k and $i,j,k+1$. The head calculated for a dry cell by equation 10c is the head that provides an outflow rate that is equal to the inflow rate to the cell, and is not the altitude of the water table in the cell. For the two-layer example illustrated in figure 3, recharge to the top cell is less than the potential vertical-flow rate through the bottom of the cell; therefore, all the recharge is added to storage in the bottom cell. The calculated head in the top cell is the value that satisfies equation 10c for constant recharge and vertical conductance. The resulting head in the top cell is equal to 8 ft for a recharge rate equal to 1 ft³/s (fig. 3).

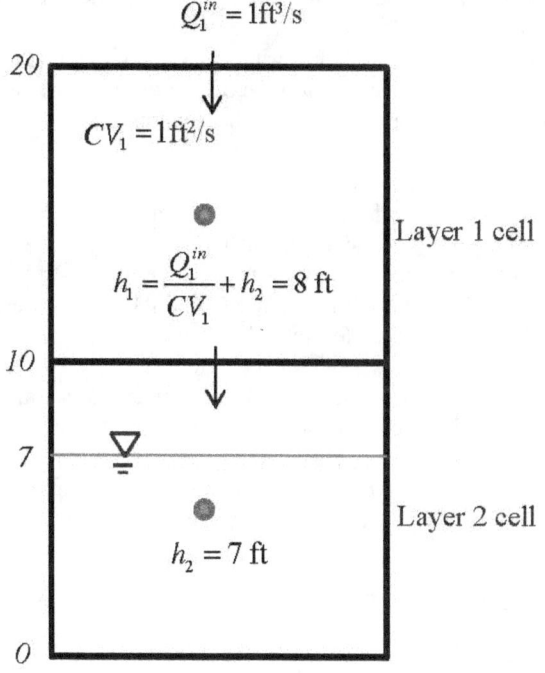

Figure 3. A two-layer model in which recharge is applied to layer 1 (cell 1) and the water table is in layer 2 (cell 2). Layer 1 is dewatered and has a head below the bottom altitude of layer 1.

MODFLOW-NWT provides two different methods for simulating dry cells in the deepest layer, and the option is defined by the NWT input file variable IBOTAV. Head for cells in the deepest layer, and for single-layer models, cannot fall below the cell-bottom altitude if the input variable IBOTAV is set to 1; otherwise, heads in the deepest layer can fall below the cell-bottom altitude. The value of IBOTAV does not affect the solution because dry bottom-layer cells are effectively ignored in the solution (that is, for a cell that does not receive inflow, the coefficients are essentially zero in the row of the matrix in which a bottom-layer cell is the diagonal element). IBOTAV is provided because it can affect convergence behavior for a cell with thin saturated thickness. The value of IBOTAV that provides the fastest convergence rate appears to be problem specific.

An important difference between simulating unconfined aquifers with MODFLOW-NWT compared to the standard MODFLOW-2005 is that groundwater heads will be calculated for dry cells, whereas standard MODFLOW-2005 excludes these calculations. Thus, it is necessary for the model user to interpret the head in a cell relative to the cell bottom. If the head in a cell is at or below the cell-bottom altitude, then the water table is not contained within this cell. If the head in the lowermost cell is at or below the cell-bottom altitude, then the aquifer is not horizontally continuous within the model domain. HDRY may be printed to output files [for example Harbaugh (2005), p. 8-26] for all cells in which the head is less than a small threshold value (such as 2 mm) above the cell bottom altitude, although internally the model uses the calculated-head values for dry cells during the simulation. Heads for dewatered cells are set to HDRY just before they are printed to the output file for dry cells and then are reset back to their calculated values prior to the following time step. If the head solution from one simulation will be used as starting heads for a subsequent simulation, or if the Observation Process is used (Harbaugh and others, 2000), then HDRY should not be printed to the output file for dry cells (that is, the UPW Package input variable should be set as IPHDRY=0).

Storage Calculation

The UPW formulation is such that a cell with head below the cell bottom has no water in storage, so changes in storage also are zero for these cells. The model accounts for this situation by setting the storage coefficient for a dry cell to zero. The Newton method requires the storage coefficient to transition smoothly with continuous derivatives. However, smoothing the storage coefficient creates the possibility for mass-balance errors to occur because the storage parameter is nonlinear for cell drying/rewetting. Mass-balance errors

are avoided with the Euler approximation (Celia and others, 1990). The groundwater-flow equation storage term is (Harbaugh, 2005; eqn. 5-33)

$$\Delta S_c = \frac{S_s \Delta z \Delta r \Delta c}{\Delta t}(h^{t+\Delta t} - h^t), \quad (11)$$

where ΔS_c is the volumetric storage change rate for confined conditions, S_s is the confined storage capacity (units of inverse length), and Δz is the cell thickness. Changes in unconfined storage are calculated according to

$$\Delta S_u = \frac{S_y \Delta z \Delta r \Delta c}{\Delta t}(Y_n^{t+\Delta t} - Y_n^t) + \frac{S_y \Delta z \Delta r \Delta c}{\Delta t}\frac{\partial Y_n}{\partial h}\Delta h^n, \quad (12)$$

where ΔS_u is the volumetric storage change rate for unconfined conditions; S_y is the specific yield; $\Delta h^n = h^n - h^{n-1}$; h^n and h^{n-1} are the groundwater head at iteration n and $n-1$; Y_n is the value of the smoothing function shown in figure 1 and defined according to the bracketed terms in equation 9. If head is below the cell bottom, then Y_n is set to an arbitrary small number (1×10^{-9}); if head is above the cell top Y_n is set equal to one. $\frac{\partial Y_n}{\partial h}$ is the slope of the curve shown in figure 1 evaluated at Y_n. For unconfined conditions, equation 11 is multiplied by Y_n to provide a smooth transition between confined storage and dry conditions of zero storage (Huyakorn and others, 1986). Thus, the total storage change over a time step is calculated by multiplying equation 11 by Y_n and adding it to equation 12. However, for confined layers, storage changes are calculated using equation 11 and no smoothing is applied.

Newton Formulation

Newton Linearization Method

Equations 9, 11, and 12, which are written for all active cells, can be substituted into equation 1 to produce a system of nonlinear algebraic equations similar to those with MODFLOW-2005. These equations can be solved using the Picard iterative-solution method for the groundwater-flow equation discussed by McDonald and Harbaugh (1988) and Harbaugh (2005). Here, however, the UPW Package is combined with the Newton linearization method in MODFLOW-NWT. The Newton method solves a system of equations that can be written in symbolic form as

$$J(h^{n-1})\Delta h^n = R^{n-1}, \quad (13)$$

where n and $n-1$ are the nonlinear iteration counters for the present and previous iterations, respectively; J is the Jacobian matrix $J_{r,l} = \frac{\partial R_r}{\partial h_l}$, and l is an index ranging from 1 to the total number of active cells starting at the upper left cell and counting along columns, then rows, and then layers; r is the index for each row in the Jacobian; $\Delta h^n = h^n - h^{n-1}$; h^n and h^{n-1} are the groundwater head at iteration n and $n-1$; R is the residual vector representing cell-by-cell errors in water balance. R is calculated by summing all cell inflows and outflows to each cell. Based on equation 1, $R_{i,j,k}$ for cell i,j,k is written:

$$
\begin{aligned}
R_{i,j,k} = {} & CV_{i,j,k-1/2}h_{i,j,k-1} + CR_{i,j-1/2,k}h_{i,j-1,k} \\
& + CC_{i-1/2,j,k}h_{i-1,j,k} + (-CV_{i,j,k-1/2} \\
& - CR_{i,j-1/2,k} - CC_{i-1/2,j,k} - CV_{i,j,k+1/2} \\
& - CR_{i,j+1/2,k} - CC_{i+1/2,j,k} + HCOF_{i,j,k})h_{i,j,k} \\
& + CV_{i,j,k+1/2}h_{i,j,k+1} + CR_{i,j+1/2,k}h_{i,j+1,k} \\
& + CC_{i+1/2,j,k}h_{i+1,j,k} - RHS_{i,j,k}.
\end{aligned} \tag{14}
$$

Equation 13 is decomposed and rearranged in terms of h^n to give (Panday and Huyakorn, 2008)

$$
J(h^{n-1})h^n = R^{n-1} + J(h^{n-1})h^{n-1}. \tag{15}
$$

Equation 15 provides an expression for the Newton linearized groundwater-flow equation in terms of the MODFLOW-2005 solution variables to maintain compatibility with the other MODFLOW packages. When rearranged, equation 15 reduces to the standard system of linear equations that is solved by MODFLOW-2005, $Ah^n = B$, for linear groundwater-flow problems, where A is the conductance matrix, and B is the source/sink term.

The terms in the Jacobian representing the groundwater-flow equation are calculated using analytical derivatives (Painter and others, 2008). The diagonal of the Jacobian ($J_{r,l(i,j,k)} = \frac{\partial R_r}{\partial h_l}$) with all h evaluated at iteration $n-1$ can be written as

$$
\begin{aligned}
J_{r,l(i,j,k)} = {} & \frac{\partial CV_{i,j,k-1/2}}{\partial h_{i,j,k}}(h_{i,j,k-1} - h_{i,j,k}) + \frac{\partial CC_{i-1/2,j,k}}{\partial h_{i,j,k}}(h_{i-1,j,k} - h_{i,j,k}) \\
& + \frac{\partial CR_{i,j-1/2,k}}{\partial h_{i,j,k}}(h_{i,j-1,k} - h_{i,j,k}) + \frac{\partial HCOF_{i,j,k}}{\partial h_{i,j,k}}h_{i,j,k} \\
& + \frac{\partial CV_{i,j,k+1/2}}{\partial h_{i,j,k}}(h_{i,j,k+1} - h_{i,j,k}) + \frac{\partial CC_{i+1/2,j,k}}{\partial h_{i,j,k}}(h_{i+1,j,k} - h_{i,j,k}) \\
& + \frac{\partial CR_{i,j+1/2,k}}{\partial h_{i,j,k}}(h_{i,j+1,k} - h_{i,j,k}) - CV_{i,j,k-1/2} - CR_{i,j-1/2,k} - CC_{i-1/2,j,k} \\
& - CV_{i,j,k+1/2} - CR_{i,j+1/2,k} - CC_{i+1/2,j,k} + HCOF_{i,j,k} - \frac{\partial RHS_{i,j,k}}{\partial h_{i,j,k}},
\end{aligned} \tag{16}
$$

where $J_{r,l(i,j,k)}$ is the value of the element at row r and column l of the Jacobian matrix corresponding to cell i,j,k, and the off-diagonals can be written as

$$J_{r,l(i,j,k-1)} = CV_{i,j,k-1/2} + \frac{\partial CV_{i,j,k-1/2}}{\partial h_{i,j,k-1}}(h_{i,j,k-1} - h_{i,j,k}), \quad (17a)$$

$$J_{r,l(i,j-1,k)} = CR_{i,j-1/2,k} + \frac{\partial CR_{i,j-1/2,k}}{\partial h_{i,j-1,k}}(h_{i,j-1,k} - h_{i,j,k}), \quad (17b)$$

$$J_{r,l(i-1,j,k)} = CC_{i-1/2,j,k} + \frac{\partial CC_{i-1/2,j,k}}{\partial h_{i-1,j,k}}(h_{i-1,j,k} - h_{i,j,k}), \quad (17c)$$

$$J_{r,l(i+1,j,k)} = CC_{i+1/2,j,k} + \frac{\partial CC_{i+1/2,j,k}}{\partial h_{i+1,j,k}}(h_{i+1,j,k} - h_{i,j,k}), \quad (17d)$$

$$J_{r,l(i,j+1,k)} = CR_{i,j+1/2,k} + \frac{\partial CR_{i,j+1/2,k}}{\partial h_{i,j+1,k}}(h_{i,j+1,k} - h_{i,j,k}), \quad (17e)$$

$$J_{r,l(i,j,k+1)} = CV_{i,j,k+1/2} + \frac{\partial CV_{i,j,k+1/2}}{\partial h_{i,j,k+1}}(h_{i,j,k+1} - h_{i,j,k}). \quad (17f)$$

The derivative of conductance terms in equations 16 and 17 can be calculated in the row direction, for example, according to

where $\dfrac{\partial CR_{i,j-1/2,k}}{\partial h_{i,j,k}}$ can only be non-zero when $h_{i,j,k}$ is greater than $h_{i,j-1,k}$ (upstream weighting). The same formulation is used for the column and the layer directions. However, vertical conductance (CV) is constant and the derivative terms in equations 16 and 17 are zero for the vertical direction.

The Newton method results in an asymmetric matrix with a main diagonal and six off-diagonal terms. For flow between two nodes (see eqn. 2), the partial derivative of the flow equation with respect to h_1 is not necessarily the same as the partial derivative of the flow equation with respect to h_2. Thus, all non-zero elements of the Jacobian are stored. Equation 16 provides the diagonal terms and equations 17a through 17f provide the off-diagonal terms for one row of the Jacobian. This results in at most a 7-point stencil for a three-dimensional finite-difference groundwater-flow problem. The Picard-equation matrix also results in a 7-point stencil structure but it is symmetric because the conductance (eqn. 3) between nodes 1 and 2 is the same as the conductance between nodes 2 and 1; hence, only the bands in the matrix corresponding to CC, CR, and CV need to be stored. The structure of the matrix produced by the Newton method is identical to that produced by previous versions of MODFLOW, and is shown by McDonald and Harbaugh (1988, fig. 46, p. 12-3). However, because the Picard method produces a matrix with symmetry among the lower and upper off-diagonal elements, MODFLOW-2005 only stores the upper off diagonals in computer memory.

$$\frac{\partial CR_{i,j-1/2,k}}{\partial h_{i,j,k}} = 0 \qquad\qquad X < 0$$

$$\frac{\partial CR_{i,j-1/2,k}}{\partial h_{i,j,k}} = \Delta C_i \Delta Z_k \frac{K_{i,j-1/2,k}}{\Delta R_{j-1}} \frac{AX}{\Omega(TOP_{up} - BOT_{up})} \qquad\qquad 0 < X \leq \Omega$$

$$\frac{\partial CR_{i,j-1/2,k}}{\partial h_{i,j,k}} = \Delta C_i \Delta Z_k \frac{K_{i,j-1/2,k}}{\Delta R_{j-1}} \frac{A}{(TOP_{up} - BOT_{up})} \qquad\qquad \Omega < X \leq 1-\Omega$$

$$\frac{\partial CR_{i,j-1/2,k}}{\partial h_{i,j,k}} = \Delta C_i \Delta Z_k \frac{K_{i,j-1/2,k}}{\Delta R_{j-1}} \left[1 - \frac{-A(1.0 - X)}{\Omega(TOP_{up} - BOT_{up})}\right] \qquad\qquad 1-\Omega < X < 1$$

$$\frac{\partial CR_{i,j-1/2,k}}{\partial h_{i,j,k}} = 0 \qquad\qquad X \geq 1, \qquad (18)$$

MODFLOW-NWT requires more than twice the amount of random-access memory (RAM) required by MODFLOW-2005 solvers to store the full matrix produced by the Newton method, and memory requirements may exceed available memory of desktop computers for very large problems (that is, those with more than 5 million cells). However, problems as large as 6 million cells have been solved using MODFLOW-NWT on a desktop computer with a 64-bit operating system. A problem with 6 million nodes required approximately 8 Gigabytes of RAM. Despite an increase in memory requirements, the unstructured linear solvers available in MODFLOW-NWT provide greater flexibility than do the structured matrix solvers, such as for coupling additional differential equations to the groundwater-flow equation.

The Newton method requires more calculations per iteration compared to the Picard method; however, the Newton method typically requires fewer iterations to find a solution due to its faster convergence properties. Relative efficiency of the Newton and Picard methods seems to be problem specific (Kuiper, 1987; Mehl, 2006). There is evidence that the Newton method performs better for highly nonlinear problems when combined with relaxation schemes and residual control (Cooley, 1983; Jacobs, 1988; Press, 2007).

Under-Relaxation Schemes

Cooley (1983) demonstrated that the Newton solution method requires under-relaxation to provide stable solutions. Under-relaxation is a method for calculating the head solution for a particular nonlinear iteration that weights the solution from previous iterations with the present iteration. Solution of equation (1) provides h for all active cells and Δh is calculated as $\Delta h = h^n - h^{n-1}$. The new head values after under-relaxation are then calculated according to

$$\Delta h_{i,j,k}^{n-1,\eta} = (1-\gamma)\Delta h_{i,j,k}^{n} + \gamma \Delta h_{i,j,k}^{n-1}, \qquad (19)$$

where $\Delta h_{i,j,k}^{n-1,\eta}$ represents changes in head weighted with the head from the previous iteration using a weighting factor γ (input variable DBDGAMMA). Head for the current iteration is then calculated from equation 19 and additional weighting:

$$h_{i,j,k}^{n,\eta} = h_{i,j,k}^{n-1} + w_{i,j,k}^{n}\Delta h_{i,j,k}^{n} + m\Delta h_{i,j,k}^{n-1,\eta}, \qquad (20)$$

where m is a constant value that is used to weight Newton solutions from previous iterations, referred to as a momentum coefficient (input variable MOMFACT). The relaxation parameter (or learning rate) $w_{i,j,k}^{n}$ is calculated in one of two ways, depending on whether the solution oscillates over nonlinear iterations. If the solution oscillates, the weighting factor is calculated as

$$w_{i,j,k}^{n} = w_{i,j,k}^{n-1} - \theta w_{i,j,k}^{n-1}; \qquad (21)$$

otherwise, while $w_{i,j,k}^{n}$ is less than one, the weighting factor is calculated as

$$w_{i,j,k}^{n} = w_{i,j,k}^{n-1} + \kappa. \qquad (22)$$

The coefficients θ ($0< \theta <1$) and κ ($0< \kappa <1$) are themselves weighting factors that are defined with the user-input variables DBDTHETA and DBDKAPPA, respectively. The under-relaxation methodology discussed above for controlling the step-size of a Newton iteration is adapted from the delta-bar-delta technique found in neural-network literature (Smith, 1993).

The Newton method can, in some cases, overshoot a solution when derivatives change abruptly as a function of h, which may then prevent convergence of MODFLOW-NWT. An option is available to use residual control with MODFLOW-NWT if the residual increases significantly. Press (2007) provides a globally convergent backtracking scheme for Newton solutions of nonlinear equations. Residual control reduces Δh between iterations by multiplying by a factor less than one until the error ceases to decrease. Residual control may occur if the user-specified flag, BACKFLAG, is greater than zero. If residual control is active, Δh is reduced according to the scheme of Press (2007):

$$\begin{aligned} &\text{if } (RMSE^n > F_r RMSE^{n-1}) \\ &\text{while } (RMSE^l < F_r RMSE^{l-1}) \\ &\text{then } \Delta h_{i,j,k}^l = B_r \Delta h_{i,j,k}^l, \end{aligned} \qquad (23)$$

where F_r is a user-specified residual-control tolerance (input variable BACTOL) and B_r is a user-specified reduction factor (input variable BACKREDUCE). Residual control should not be used (BACKFLAG=0) unless MODFLOW-NWT is having trouble converging with under-relaxation. A residual-control iteration differs from a standard nonlinear iteration because the Jacobian is not assembled and the linear equations are not solved; however, the sequence of calls within the MODFLOW-NWT main source file are carried out during a residual-control iteration to recalculate the residuals for the reduced Δh. For a residual-control iteration, the number of inner iterations will be zero.

Example Input for the NWT Input File

The NWT input file requires specification of more input variables than do the other currently available MODFLOW-2005 solvers. These extra input variables can provide flexibility for achieving convergence. Table 2 provides suggested values that can be used as a starting point for the input variables. However, optimal value for each of these variables is usually problem specific and determined by trial and error.

Table 2. Suggested input values for the NWT input file.

[Dashes (–) indicate that values are not applicable]

Input variable name	Default values			Range
	Iteration Control			
HEADTOL	1×10^{-4} (L)*			–
FLUXTOL	500 (L^3/T)*			–
MAXITEROUT	100			10–2,000
	Dry Cell Tolerance			
THICKFACT	0.00001			1×10^{-6}-1
	NWT Options			
LINMETH	1			1 or 2
IPRNWT	0			0 or 2
IBOTAV	0**			0 or 1
	Under relaxation Input			
	Model Complexity			
	Simple	Moderate	Complex	
DBDTHETA	0.97	0.7	0.4	0–1
DBDKAPPA	0.0001	0.0001	0.00001	0–1
DBDGAMMA	0.0	0.0	0.0	0–1
MOMFACT	0.0	0.1	0.1	0–1
	Residual Control			
	Model Complexity			
	Simple	Moderate	Complex	
BACKFLAG	0***	0	0	0 or 1
MAXBACKITER	–	–	–	1–100
BACKTOL	–	–	–	1–2
BACKREDUCE	–	–	–	0.05–0.99
	Linear Solution Control and Options for GMRES			
	Model Complexity			
	Simple	Moderate	Complex	
MAXITINNER	50	50	50	25–1,000
ILUMETHOD	2	2	2	1 or 2
LEVFILL	1	1	1	5–10 for ILUMETHOD=1 0–2 for ILUMETHOD=2
STOPTOL	1×10^{-10}	1×10^{-10}	1×10^{-10}	1×10^{-8}-1×10^{-12}
MSDR	5	10	15	5–20

Table 2. Suggested input values for the NWT input file.—Continued

[Dashes (–) indicate that values are not applicable]

Input variable name	Default values			Range
	Linear Solution Control and Options for χMD			
	Model Complexity			
	Simple	Moderate	Complex	
IACL	2	2	2	0,1, or 2
NORDER	1	1	0	0,1, or 2
LEVEL	0	1	3	1–10
NORTH	2	2	7	4-10 for Orthomin 2 otherwise
IREDSYS	0	0	0	0 or 1
RRCTOLS	0.0	0.0	0.0	0–0.0001
IDROPTOL	1	1	1	0 or 1
EPSRN	1×10^{-3}	1×10^{-3}	1×10^{-4}	5×10^{-5}-1×10^{-3}
HCLOSEXMD	1×10^{-4}	1×10^{-4}	1×10^{-4}	1×10^{-3}-1×10^{-5}
MXITERXMD	50	50	50	25–1,000

[*]These values are dependent on the units specified in the MODFLOW-2005 discretization input file. Values given are for units of meters and days.

[**]The optimal value for IBOTAV is problem specific. Values of 0 and 1 should be tested for each problem.

[***]BACKFLAG should be set to 0 (residual control set to inactive) unless there are convergence problems. "OPTIONS" must be set to "SPECIFIED" if the residual control option is used.

Default values for the head- and flux-convergence tolerances (HEADTOL and FLUXTOL) are dependent on the specified length units and therefore should be adjusted based on the units used. In the example shown in table 2, meters were used for length units and days were used for time units. Convergence is reached when "Maximum-Head-Change" is less than or equal to HEADTOL and "RMS" is less than or equal to FLUXTOL. "Maximum-Head-Change" and "RMS" (or "RMS-New" if BACKFLAG is greater than zero) are printed to the main Listing file at the end of each iteration when IPRNWT is set greater than 0. Refer to Appendix A, section "MODFLOW-NWT Listing File" for more details.

Values shown in table 2 are representative of the optimal values for the problems that were tested for the development of MODFLOW-NWT. Optimal values were not the same for all problems tested. For simulations experiencing problems with convergence, DBDTHETA and DBDKAPPA will likely need adjustment. Solver testing revealed THICKFACT (variable Ω in eqn. 9) is optimal between 1×10^{-6} to 1×10^{-4}, depending on the problem. However, a value of 1×10^{-5} was optimal for most problems tested.

Default values for the input variables for MODFLOW-NWT are available for three separate simulation cases, and are specified by keywords for variable OPTIONS: (1) nearly linear models (keyword SIMPLE), (2) moderately nonlinear models (MODERATE), and (3) strongly nonlinear, complex models (COMPLEX). A fourth option is to specify values for each solver input variable (keyword SPECIFIED). The default values provide a good starting point and are appropriate for a wide range of model simulations. The input variable "OPTIONS" must be set to "SPECIFIED" when using the residual-control (backtracking) option.

The initial head distribution specified for a simulation is a critical factor that affects model computation time. Significant computational time can be saved by using the steady-state head solution with initial hydraulic properties for the starting transient head distribution. This is especially true for models used with automated parameter estimation. If the steady-state solution from a simulation is used for the starting transient head distribution, then the option of setting heads to HDRY for cells with heads below the cell bottom should not be used for the steady-state simulation.

Additional MODFLOW Packages Modified for MODFLOW-NWT

Several packages require terms to be added to the right-hand-side vector (RHS_{bc}) and conductance matrix ($HCOF_{bc}$) in MODFLOW-NWT. These packages do not require special treatment for MODFLOW-NWT. However, nonlinear stress packages also benefit from the stronger convergence properties of the Newton method. Changes therefore were made to some of the existing MODFLOW stress packages to calculate the stress derivative with respect to head and add them appropriately to the Jacobian matrix. The WEL Package, the Streamflow-Routing Package (SFR2; Niswonger and Prudic, 2005), and the Unsaturated-Zone Flow Package (UZF1; Niswonger and others, 2006) are solved using the Newton method, and the derivative of groundwater inflow or outflow as a function of groundwater head are calculated in the Formulate subroutines of these packages. Other nonlinear packages also could be modified such that they also can benefit from the stronger convergance properties of the Newton method. Appendix B presents a simple example illustrating how to modify a nonlinear stress package for adding derivatives to the Jacobian in MODFLOW-NWT. If a nonlinear stress package is not modified to add the stress derivative to the Jacobian matrix solved by MODFLOW-NWT, then it is linearized using the Picard method, whereas the groundwater-flow equation is linearized using the Newton method.

It can be problematic to calculate derivatives of nonlinear boundary conditions for some of the nonlinear MODFLOW-2005 packages. A good example is the LAK Package, in which a single lake interacts with multiple groundwater cells. Consequently, for such packages, the Picard method is used and $HCOF_{bc}$ and RHS_{bc} are the head-dependent and non-head-dependent terms of the boundary condition, respectively, which is the same method used by MODFLOW-2005.

Well Package

Negative pumping rates specified in the Well Package are reduced to zero when the groundwater head drops to the cell bottom using a cubic formula and its derivative. This option is only available for unconfined (convertible) layers. This formula decreases the pumping rate as the head drops below a user-specified percentage of the cell thickness. For drawdown conditions, the following equations were added to the Formulate and Budget routines within the WEL Package:

$$IF\ Q_{Wel} < 0$$

$$Q_{net} = Q_{Wel}\, x^2 \left(\frac{-2}{\zeta^3} x + \frac{3}{\zeta^2} \right) \qquad 0 < x < \zeta$$

$$\frac{\partial Q_{net}}{\partial h} = Q_{Wel} \left(\frac{6}{\zeta^2} x - \frac{6}{\zeta^3} x^2 \right) \qquad 0 < x < \zeta$$

$$Q_{net} = Q_{Wel}, \quad \frac{\partial Q_{net}}{\partial h} = 0 \qquad x \geq \zeta$$

$$Q_{net} = 0, \quad \frac{\partial Q_{net}}{\partial h} = 0 \qquad x \leq 0, \qquad (24)$$

and

$$RHS = RHS - Q_{net} \qquad 0 < x < \zeta$$

$$\frac{\partial RHS}{\partial h} = \frac{\partial RHS}{\partial h} - \frac{\partial Q_{net}}{\partial h} \qquad 0 < x < \zeta$$

$$RHS = RHS - Q_{Wel} \qquad x \geq \zeta, \qquad (25)$$

where $x = (h - BOT)$, $\zeta = \Phi(TOP - BOT)$, Φ ranges between 0–1 and typically is a small value, such as 0.25 or smaller, Q_{wel} is the specified pumping rate, and Q_{net} is the applied pumping rate. Figure 4 shows an example of the smoothing function used to reduce negative pumping rates to zero. If the keyword SPECIFY is specified, then a value of Φ (variable PHIRAMP) must be specified following the keyword SPECIFY, and the options discussed above are implemented. If a negative pumping rate is specified in the WEL Package and the rate is limited by the amount of water in a cell, then the cell is reported in the Listing File with the reduced pumping rate.

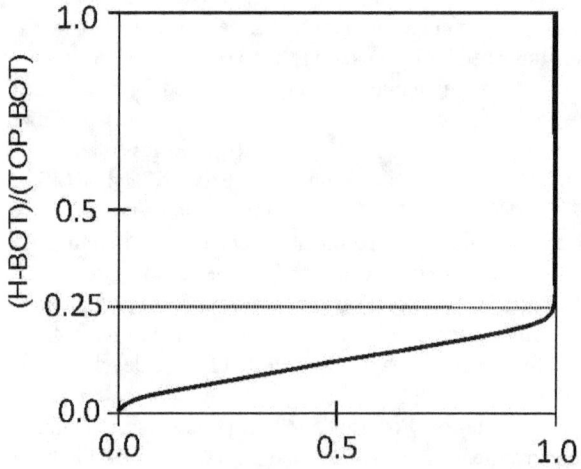

Figure 4. Curve used to smoothly reduce specified pumping to zero when cell dewaters, $\Phi = 0.25$.

Streamflow-Routing Package

Implementation of the Streamflow-Routing (SFR2) Package in MODFLOW-NWT requires calculation of $\frac{\partial HCOF_{SFR}}{\partial h}$ and $\frac{\partial RHS_{SFR}}{\partial h}$, which are added to equation 16. These values are obtained within the Formulate routine of SFR2 using the following numerical approximation of the derivative

$$\frac{\partial Q_{SFR}}{\partial h} = \frac{Q_{h+delh} - Q_h}{delh}, \qquad (26)$$

where Q_{h+delh} and Q_h are streambed seepage calculated for groundwater head h and $h + delh$, respectively, and $delh$ is a small value, such as 1.0×10^{-7}. Additionally, for unconfined (convertible) layers, a check was added to ensure that the streambed-bottom altitude (variable STRTHICK subtracted from STRTOP, Niswonger and Prudic, 2005, p. 21) is not specified less than the cell-bottom altitude to which it is assigned. If this occurs, an error message is printed to the Listing File and the model stops. Users of SFR2 do not need to modify the input file for use with MODFLOW-NWT.

Unsaturated-Zone Flow Package

Implementation of the Unsaturated-Zone Flow Package in the NWT Solver requires calculation of $\frac{\partial HCOF_{UZF}}{\partial h}$ and $\frac{\partial RHS_{UZF}}{\partial h}$, which is done using the same approach (eqn. 26) as described for the SFR2 Package. Users of UZF1 do not need to modify the input file for use with MODFLOW-NWT.

Example Problems

The following section presents three different example simulations that demonstrate some of the features and capabilities of MODFLOW-NWT. Problem 1 consists of a simple problem that provides a comparison to analytical solutions of groundwater flow in an unconfined aquifer. The second problem compares MODFLOW-NWT to MODFLOW-2005 for the three-dimensional groundwater-flow problem provided by McDonald and others (1991) that involves wetting and drying of model cells caused by recharge from an infiltration basin. Problem 3 is a hypothetical unconfined aquifer simulation. This problem includes many of the conditions that are difficult for MODFLOW-2005 to simulate and provides a good example of the enhanced convergence capabilities of MODFLOW-NWT for simulating flow in unconfined aquifers.

Problem 1—Comparison to Analytical Solutions

This example problem of one-dimensional groundwater flow was designed to compare MODFLOW-NWT model results with two analytical solutions of unconfined groundwater flow. Results also were compared with a standard MODFLOW-2005 simulation using the LPF and PCG7 Packages.

The analytical solution for Example Problem 1 is steady, unidirectional flow in an unconfined aquifer (Jacob, 1950, p. 378; Todd and Mays, 2005, p. 147). This solution is based on the assumption of a horizontal bottom and flow between two constant-head boundaries. The analytical equation for this problem is (Fetter, 1994, eqn. 5-72)

$$h = \sqrt{h_1^2 - \frac{(h_1^2 - h_2^2)x}{L}}. \qquad (27)$$

The model consisted of a grid of 100 columns, 1 row, and 1 layer; a bottom altitude of the aquifer of 0 m; and constant heads of 10 m at one end of the model and 50 m at the other end of the model. Grid cells were 50 m wide and K was 50 m/d. Table 3 lists the results for the two MODFLOW models and the analytical solution for the simulated conditions. The solution with MODFLOW-NWT is close to that of the LPF Package and the analytical solution, where errors for MODFLOW-NWT are less than 1 percent for heads and flows at all locations (table 3 and fig. 5). Errors for MODFLOW-NWT are caused by the upstream weighting formulation.

The second analytical solution represents steady flow in response to recharge in a one-dimensional system. The solution is documented in Todd and Mays (2005, p. 149). This is an unconfined system with a uniform recharge rate, a horizontal bottom, and flow between a no-flow boundary and a constant-head boundary. MODFLOW models cannot match the analytical solution exactly because they do not allow recharge to constant-head cells. Constant-head cells were made very thin (0.1 m) in the direction of flow to minimize the effect of recharge applied to them. The analytical equation for this problem can be written as (Todd and Mays, 2005)

$$h^2 = h_a^2 + \frac{W}{K}(a^2 - x^2), \qquad (28)$$

where W is the recharge rate, K is the hydraulic conductivity in the horizontal direction, h_a is the specified head at the left boundary, a is the recharge area, and x is the distance from the no-flow boundary. Similar to the first analytical comparison, the model consisted of a grid of 100 columns, 1 row, and 1 layer; a bottom altitude of 0 m; constant head of 10 m; a recharge rate of 0.001 m/d; and a horizontal hydraulic conductivity rate of 50 m/d. The discretization is 0.1 m in the row direction for the constant-head cell and 50 m for all other cells.

Table 3.   Comparisons between MODFLOW-2005 and MODFLOW-NWT solutions for a one-dimensional unconfined aquifer between constant-head boundaries.

[**Abbreviation:** m^3/d, cubic meter per day]

Node No.	Distance from constant head node (meters)	Calculated head (meters)		
		Analytical solution	MODFLOW-2005	MODFLOW-NWT
1	0	10.00	10.00	10.00
11	500	18.50	18.51	18.37
21	1,000	24.18	24.18	24.05
31	1,500	28.76	28.76	28.65
41	2,000	32.70	32.70	32.61
51	2,500	36.22	36.22	36.15
61	3,000	39.42	39.42	39.37
71	3,500	42.39	42.39	42.35
81	4,000	45.15	45.16	45.13
91	4,500	47.76	47.76	47.76
100	4,950	50.00	50.00	50.00
Total flow (m^3/d)		606.06	605.97	611.04

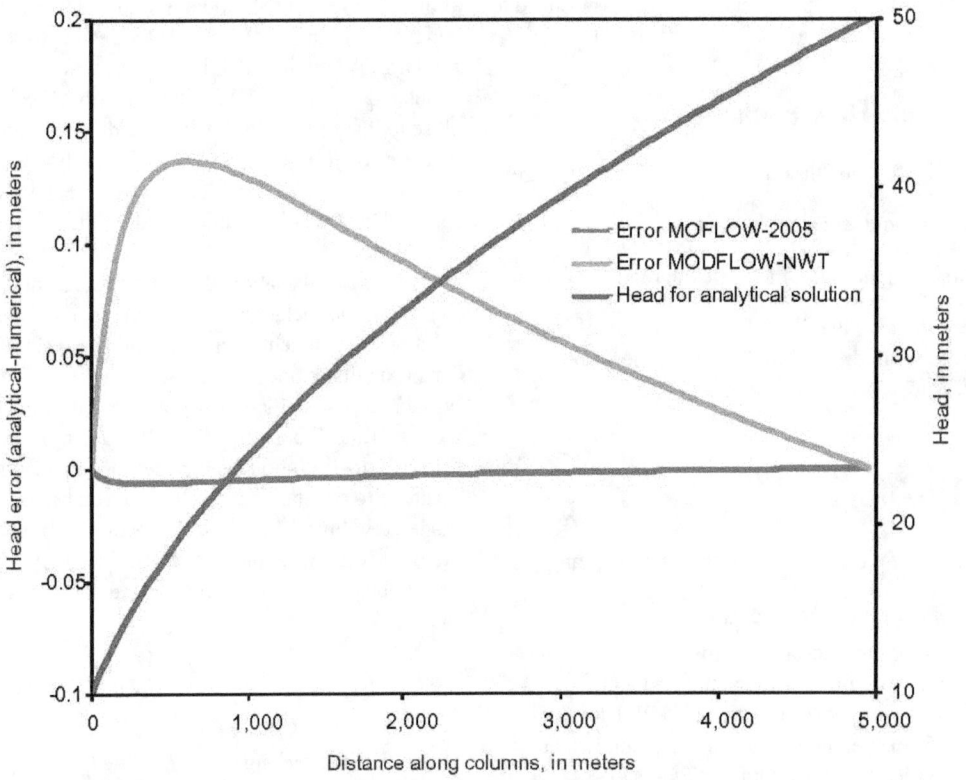

Figure 5.   Difference in head for MODFLOW-2005 and MODFLOW-NWT relative to analytical solution for unconfined flow between two constant-head boundary conditions.

Table 4 and figure 6 present the results for the two MODFLOW models and the analytical solution. The solution with MODFLOW-NWT is close to that of the LPF Package and the analytical solution, where errors for MODFLOW-NWT are less than 1 percent for heads and flows at all locations. However, the errors for MODFLOW-2005 are smaller than for MODFLOW-NWT because of the use of upstream weighting in MODFLOW-NWT, which introduces a small error into the conductance functions (fig. 1).

Table 4. Comparisons between MODFLOW-2005 and MODFLOW-NWT solutions for a one-dimensional unconfined aquifer with a recharge rate of 1×10^{-3} meters per day.

Node No.	Distance from constant head node (meters)	Calculated head (meters)		
		Analytical solution	MODFLOW-2005	MODFLOW-NWT
1	0	10.00	10.00	10.00
11	475	13.77	13.77	13.72
21	975	16.55	16.55	16.49
31	1,475	18.67	18.67	18.60
41	1,975	20.32	20.32	20.26
51	2,475	21.62	21.62	21.56
61	2,975	22.63	22.63	22.56
71	3,475	23.38	23.38	23.31
81	3,975	23.90	23.90	23.83
91	4,475	24.20	24.20	24.14
100	4,925	24.29	24.29	24.23

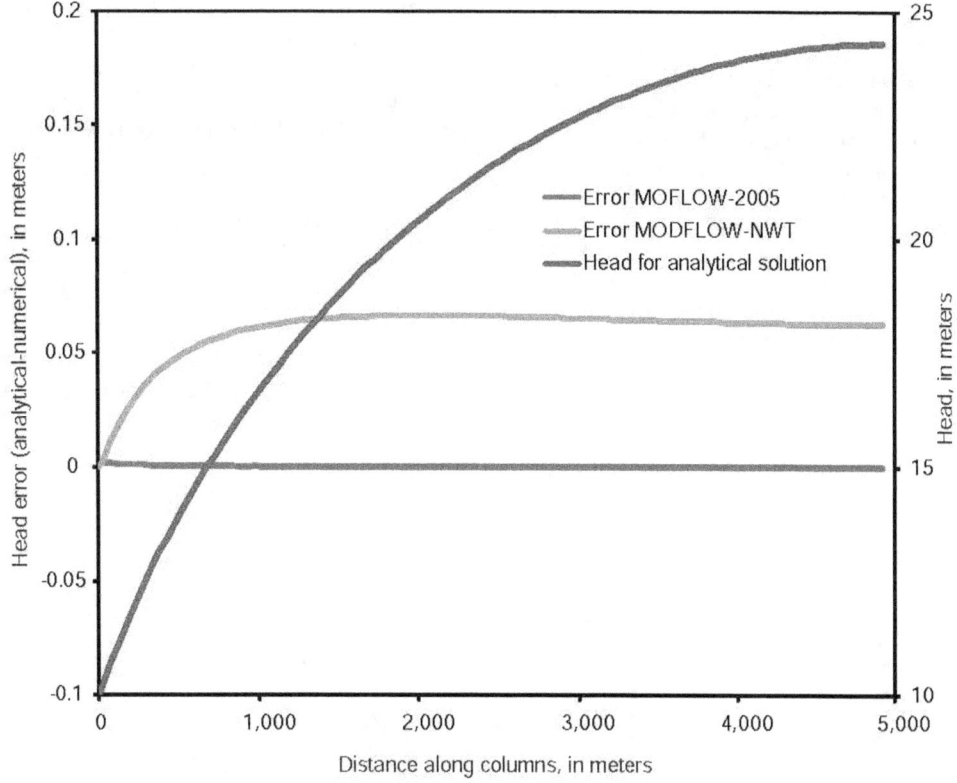

Figure 6. Difference in head for MODFLOW-2005 and MODFLOW-NWT relative to analytical solution for unconfined flow with recharge.

Problem 2—Simulation of a Water-Table Mound Resulting from Local Recharge

A second example problem is provided for three-dimensional groundwater flow. In this second example, MODFLOW-NWT is compared with the Pre-Conditioned Conjugate Gradient (PCG) Solver with the LPF Package of MODFLOW-2005. The problem was taken from McDonald and others (1991), where it is referred to as "problem 2". English units were used for this problem, which is consistent with the units used in McDonald and others (1991). This problem was chosen because it considers a more complicated groundwater-flow system that cannot be solved analytically. The problem includes a 14-layer model with the water table intersecting multiple model layers. The upper 9 layers were allowed to convert between confined and unconfined in the simulations in the LPF Package or the UPW Package. Wetting-threshold values of 0.5 ft were used for all convertible layers with the LPF Package.

The problem consists of transient recharge from a small leaky pond to a water-table aquifer. At the start of the simulation, the pond is dry. Shortly afterwards, the pond fills and begins to leak water from the bottom. The recharge subsequently causes a groundwater mound to form beneath the pond.

The simulation represents a rectangular, unconfined aquifer with a deep water table (fig. 7). The lengths of the model domain for both the MODFLOW and MODFLOW-NWT models are the same. All 14-layers in the model are 5,000 ft in the x and y directions. Each model layer has 40 rows (x-direction) and 40 columns (y-direction). All model cells are square in plan view with side dimensions of 125 ft in both the x and y directions. Layer 1 is 15 ft thick. Layers 2 through 14 are 5 ft thick. The model uses symmetry to simplify the problem by simulating one-quarter of the pond (fig. 7). The horizontal hydraulic conductivity is 5 ft/d and vertical hydraulic conductivity is 0.25 ft/d. The specific yield is 20 percent. The pond area above the aquifer is approximately 6 acres. Pond leakage is 12,500 ft³/d (65 gal/min). The water table is flat prior to the creation of the recharge pond, and the flat water table is established using a uniform constant-head boundary that surrounds the aquifer (fig. 7).

Water-table altitudes were compared for four simulation times: 190 days; 708 days; 2,630 days; and at steady state (fig. 8). Water-table altitudes were very similar for the two solutions, with a maximum difference in head of 1.08 ft directly beneath the pond for the steady-state solution. Generally, heads for MODFLOW-NWT were within 0.1 ft of the LPF Package, with heads typically lower for MODFLOW-NWT (table 5).

These minor differences indicate that the two simulations are essentially equal. Both models provide reasonable finite-difference solutions for three-dimensional groundwater flow in the unconfined system. The main advantage of MODFLOW-NWT solution is that all model cells remain active throughout the simulation and there is no need to remove or add cells during drying and rewetting. However, because there are many cells that are dewatered, MODFLOW-NWT solution required more calculations relative to MODFLOW-2005, which only makes calculations for cells that are partially or fully saturated. Consequently, MODFLOW-NWT required about three times the amount of simulation time to solve this problem.

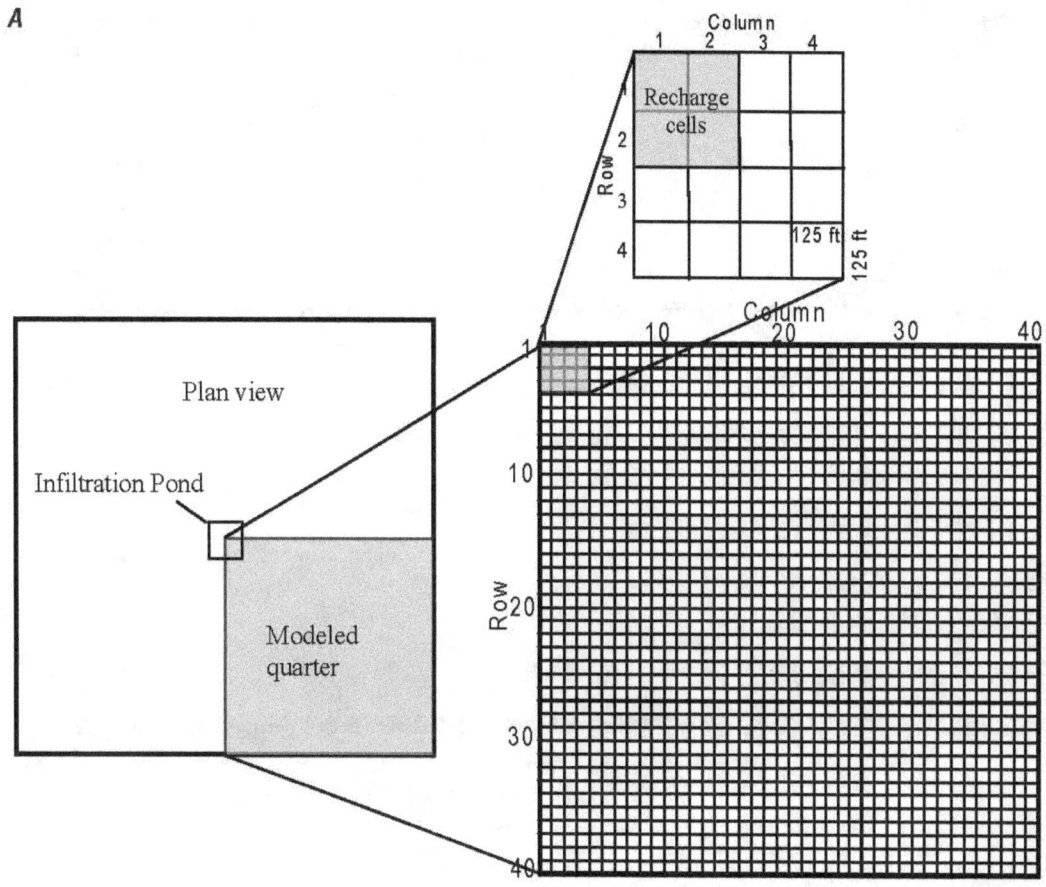

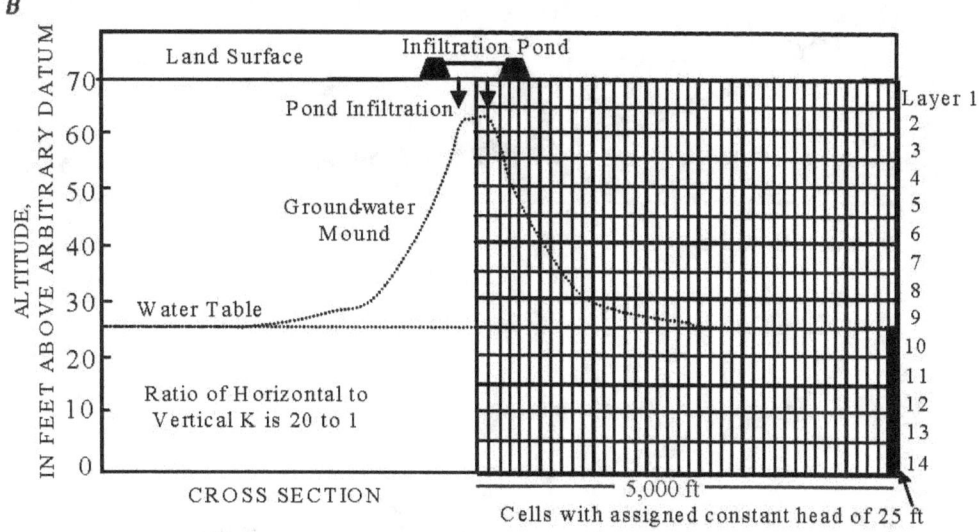

Figure 7. Diagrams showing (A) plan view of model, and (B) cross-section of model used for model comparison.

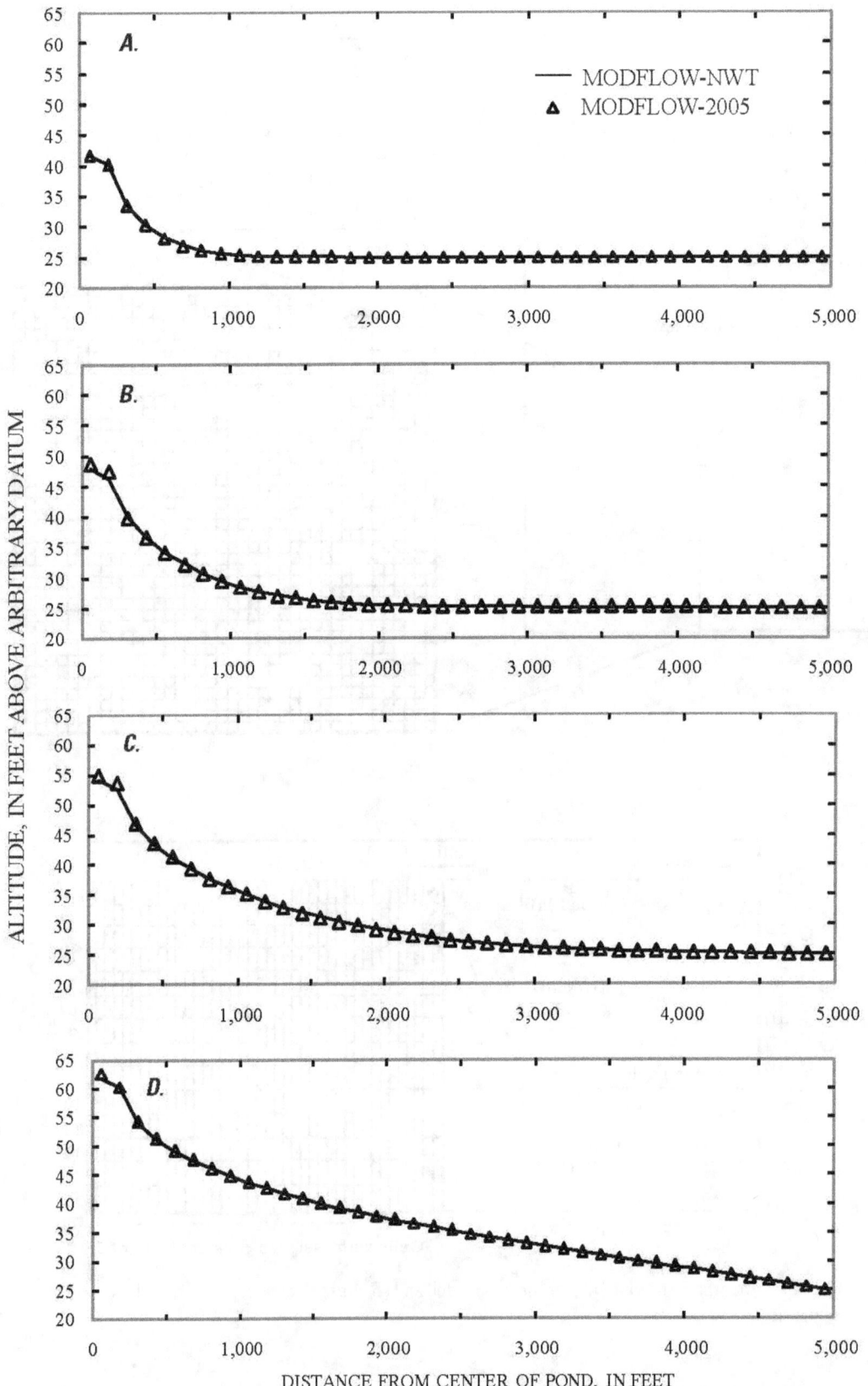

Figure 8. Comparison of water-table altitudes simulated by MODFLOW-2005 and MODFLOW-NWT. Water-table altitudes are shown for (A) 190 days, (B) 708 days, (C) 2,630 days, and (D) at steady state.

Table 5.   Comparison of simulated heads calculated by MODFLOW-2005 and MODFLOW-NWT for problem 2 of McDonald and others (1991) at selected distances from infiltration basin.

[Head values are in units of feet]

	MODFLOW-2005	MODFLOW-NWT
62.5 feet (distance from infiltration basin)		
190 days	41.6	41.44
708 days	48.76	47.62
2,630 days	54.88	53.80
Steady state	62.61	61.58
1,062.5 feet		
190 days	25.34	25.36
708 days	28.37	28.41
2,630 days	34.92	34.84
Steady state	43.72	43.59
2,062.5 feet		
190 days	25	25
708 days	25.4	25.38
2,630 days	28.58	28.55
Steady state	37.15	37.07
3,062.5 feet		
190 days	25	25
708 days	25.15	25.04
2,630 days	26.06	26.05
Steady state	32.52	32.47

Problem 3—Hypothetical Unconfined Aquifer

Example Problem 3 demonstrates the use of MODFLOW-NWT applied to a hypothetical steady-state, unconfined groundwater-flow problem. This problem is challenging to solve using MODFLOW-2005 solvers that rely on the Picard-linearization method because the aquifer-bottom altitude varies across the domain, and regions of the aquifer become dry for low recharge rates. Standard MODFLOW-2005 solvers have difficulty with this problem because recharge cannot be applied where cells become dry, and the steady-state solution is dependent on the initial heads used in the model. A comparison first is made between MODFLOW-2005 and MODFLOW-NWT with a large areal recharge rate. A second simulation then uses a lower recharge rate, which results in discontinuous regions of the aquifer that cause convergence failure in MODFLOW-2005 and very poor mass-balance error (greater than 100 percent). However, MODFLOW-NWT provides solutions for both recharge rates and, therefore, this problem provides a good example of the utility of MODFLOW-NWT for solving problems with wetting and drying of cells.

The model consists of a thin, unconfined aquifer that has a bottom altitude that ranges from about 4 to 80 m (fig. 9). The model consists of 80 columns, 80 rows, and 1 layer, and the length of cells in the row and column directions are equal to 100 m. The top altitude of the model is 200 m to avoid the water table from exceeding the top of the model. Horizontal hydraulic conductivity is 1 m/d. There are constant-head boundaries at three cells where the bottom altitude is the lowest (fig. 9). Recharge is distributed according to the bottom-altitude distribution, where larger recharge rates correspond to higher altitudes (fig. 10). Initial heads in the model are 20 m above the cell bottom-altitudes except at the constant-head cells, where the initial heads are 24 m. The simulation consists of a single steady-state stress period. Table 6 shows NWT input variables for this problem. Rewetting is active in the MODFLOW-2005 simulation, whereas rewetting is not an option in MODFLOW-NWT because rewetting occurs by default and rewetting parameters are not required.

MODFLOW-2005 and MODFLOW-NWT provide reasonable solutions with good mass-balance error for the simulation with large recharge rates (figs. 11 and 12; table 7). Figure 11 shows dry cells as dark blue. The simulated saturated thickness ranges between zero and 25 m for both MODFLOW-2005 and MODFLOW-NWT (fig. 13). MODFLOW-NWT prints the value of HDRY for heads that are less than 2 mm above the cell bottom (if UPW input variable IPHDRY is set to 1).

MODFLOW-2005 provides a solution to this problem that is very sensitive to the initial heads. The total recharge for the MODFLOW-2005 simulation is 223.6 m³/d, which is less

Table 6.   NWT input file values used for problem 3.

Input variable name	Input value
HEADTOL	0.001 (L)
FLUXTOL	1.0 (L³/T)
MAXITEROUT	500
THICKFACT	0.000001
LINMETH	2 (χMD)
IPRNWT	1
IBOTAV	1
OPTIONS	Specified
DBDTHETA	0.9
DBDKAPPA	0.0001
DBDGAMMA	0.0
MOMFACT	0.1
BACKFLAG	0
IACL	2
NORDER	0
LEVEL	3
NORTH	7
IREDSYS	0
RRCTOLS	0.0
IDROPTOL	1
EPSRN	1×10^{-4}
HCLOSEXMD	1×10^{-4}
MXITERXMD	200

than the total applied recharge of 296.9 m³/d because inactive cells reject recharge in MODFLOW-2005 (fig. 12). The larger amount of recharge in the MODFLOW-NWT simulation likely results in greater groundwater heads throughout the model domain (fig. 11).

Problem 3 was used again to simulate more arid conditions by reducing the recharge rate by three orders of magnitude (fig. 10). MODFLOW-2005 does not provide a solution for these recharge rates, and only results produced by MODFLOW-NWT are presented for this simulation. MODFLOW-NWT provides a solution for the reduced recharge rates with good mass balance, further illustrating its greater robustness for solving unconfined groundwater problems relative to MODFLOW-2005. Figures 14 and 15 show groundwater heads and saturated thicknesses for this second simulation. Only a small portion of the model domain has heads that are significantly above the cell bottom. However, all cells in the simulation have some saturated thickness (greater than 2 mm) in order to allow the applied recharge to flow horizontally toward the constant-head boundaries at the outlet of the aquifer.

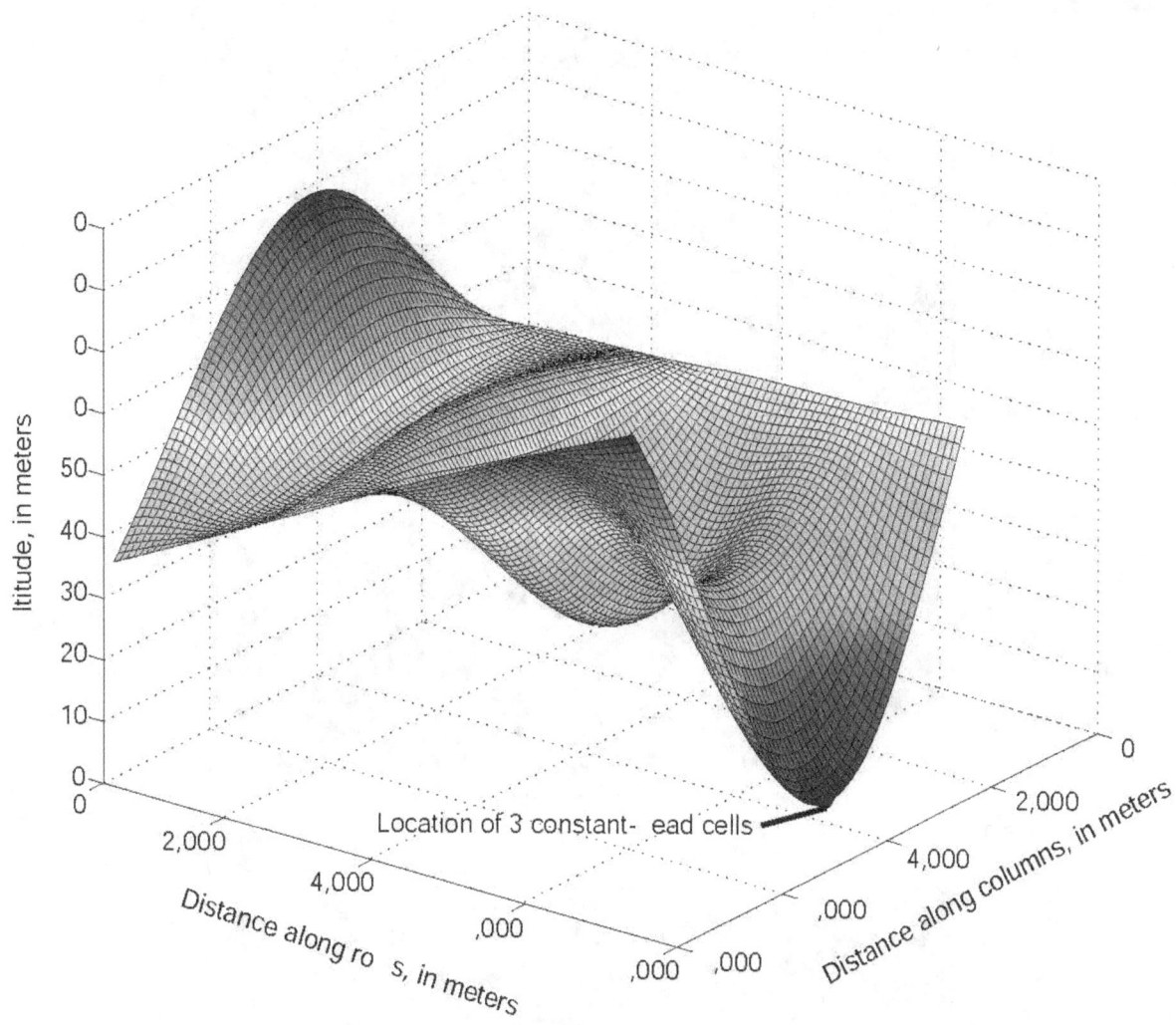

Figure 9. Distribution of layer-bottom altitudes for problem 3.

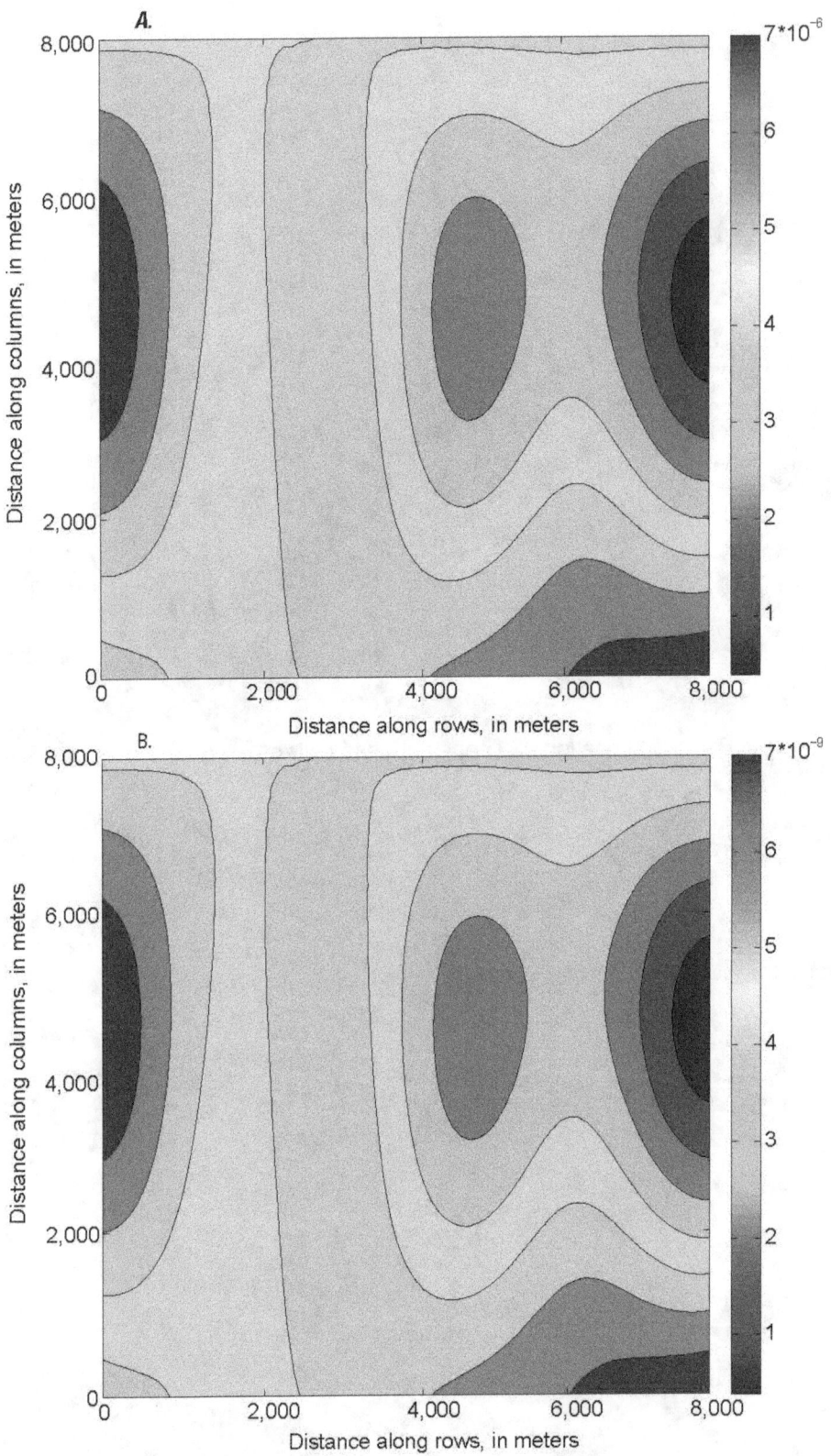

Figure 10.   Distributions of recharge used for problem 3. (*A*) higher recharge rates, and (*B*) lower recharge rates. Units for recharge are meters per day.

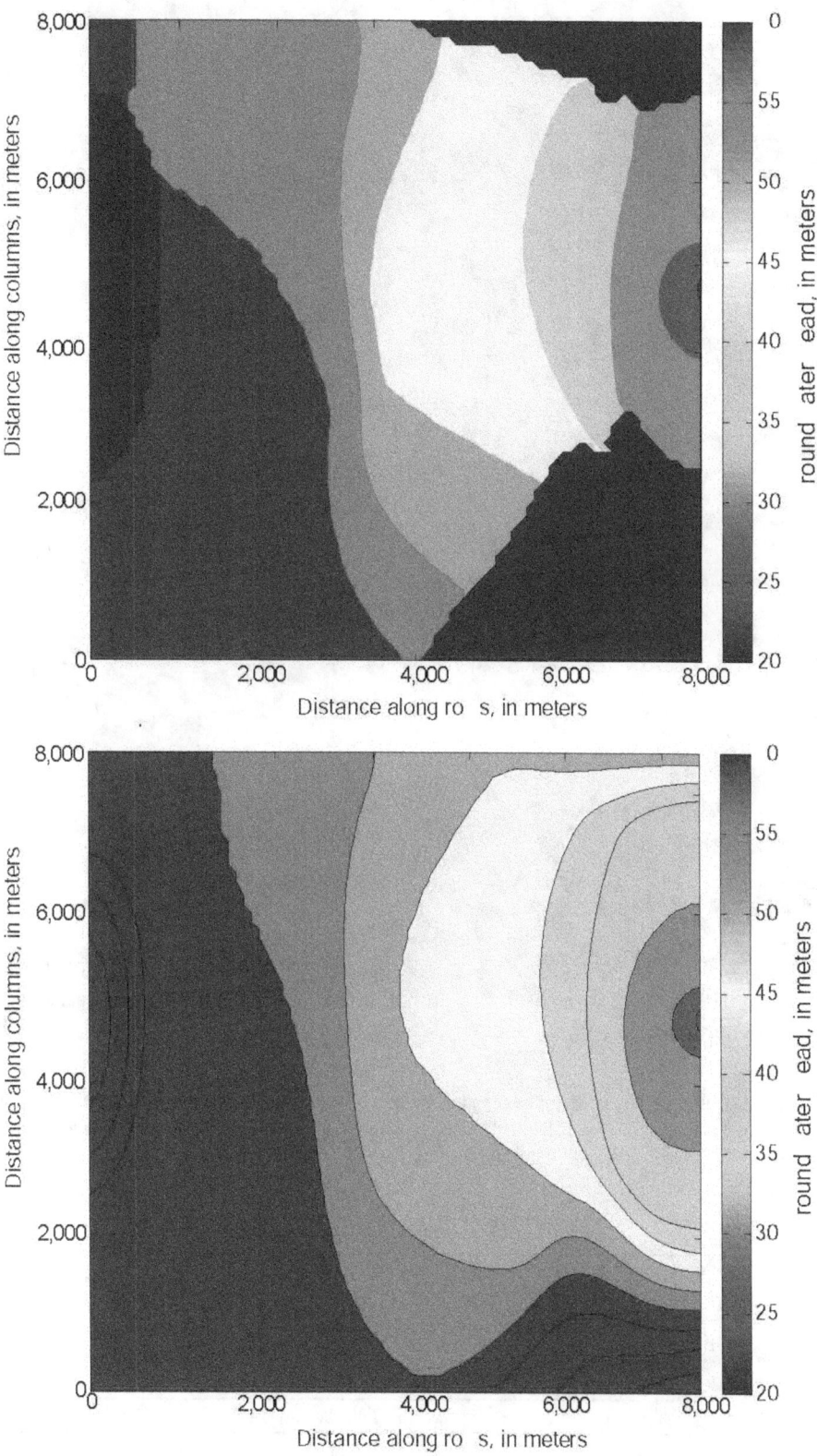

Figure 11. Distribution of heads for problem 3 with higher recharge values (fig. 10), (*A*) MODFLOW-2005, and (*B*) MODFLOW-NWT. The dark blue regions in (*A*) indicate cells that are dry in the MODFLOW-2005 simulation, whereas all cells have heads greater than 2 mm above the cell bottom in the MODFLOW-NWT simulation.

A.

VOLUMETRIC BUDGET FOR ENTIRE MODEL AT END OF TIME STEP 1 IN STRESS PERIOD 1

CUMULATIVE VOLUMES   L**3   RATES FOR THIS TIME STEP   L**3/T
-----------------   ----------------------

```
   IN:                      IN:
   ---                      ---
     STORAGE =      0.0000    STORAGE =        0.0000
     CONSTANT HEAD = 0.0000   CONSTANT HEAD =  0.0000
     RECHARGE =   81609.38    RECHARGE =     223.58

     TOTAL IN =   81609.38    TOTAL IN =     223.58

   OUT:                     OUT:
   ----                     ----
     STORAGE =      0.0000    STORAGE =        0.0000
     CONSTANT HEAD = 81727.92 CONSTANT HEAD = 224.02
     RECHARGE =     0.0000    RECHARGE =       0.0000

     TOTAL OUT =  81727.92    TOTAL OUT =    224.02

   IN - OUT =    -161.2578    IN - OUT =      -0.4418

   PERCENT DISCREPANCY =   -0.20   PERCENT DISCREPANCY =   -0.20
```

B.

VOLUMETRIC BUDGET FOR ENTIRE MODEL AT END OF TIME STEP 1 IN STRESS PERIOD 1

CUMULATIVE VOLUMES   L**3   RATES FOR THIS TIME STEP   L**3/T
-----------------   ----------------------

```
   IN:                      IN:
   ---                      ---
     STORAGE =      0.0000    STORAGE =        0.0000
     CONSTANT HEAD = 0.0000   CONSTANT HEAD =  0.0000
     RECHARGE =   108351.14   RECHARGE =     296.8524

     TOTAL IN =   108351.14   TOTAL IN =     296.8524

   OUT:                     OUT:
   ----                     ----
     STORAGE =      0.0000    STORAGE =        0.0000
     CONSTANT HEAD = 108353.46 CONSTANT HEAD = 296.8588
     RECHARGE =     0.0000    RECHARGE =       0.0000

     TOTAL OUT =  108353.46   TOTAL OUT =    296.8588

   IN - OUT =    -2.3203      IN - OUT =     -6.3477E-03

   PERCENT DISCREPANCY =    0.00   PERCENT DISCREPANCY =    0.00
```

Figure 12. Budget reports for (*A*) MODFLOW-2005 and (*B*) MODFLOW-NWT for higher recharge values (fig. 10).

Table 7. Simulated groundwater heads in selected cells for problem 3 and the higher recharge rates.
[Columns and rows are numbered starting in the upper left corner of the model domain, where columns increase from left to right and rows increase from top to bottom]

	Column 1	Column 20	Column 40	Column 60	Column 80
	MODFLOW-2005 (meters)				
Row 1	61.9	61.6	59.9	dry	dry
Row 20	62	61.3	54.2	dry	dry
Row 40	dry	60.6	49.3	44.6	29.5
Row 60	dry	59.4	49.5	43.9	32.6
Row 80	60	58.8	53.4	dry	dry
	MODFLOW-NWT (meters)				
Row 1	62.5	62.2	60.5	69.2	79.4
Row 20	62.7	61.9	55.8	54	41.8
Row 40	77	61.3	50.9	45.8	31.7
Row 60	72.6	60.2	50.9	45	34.6
Row 80	61	59.7	54.2	51.9	53.3

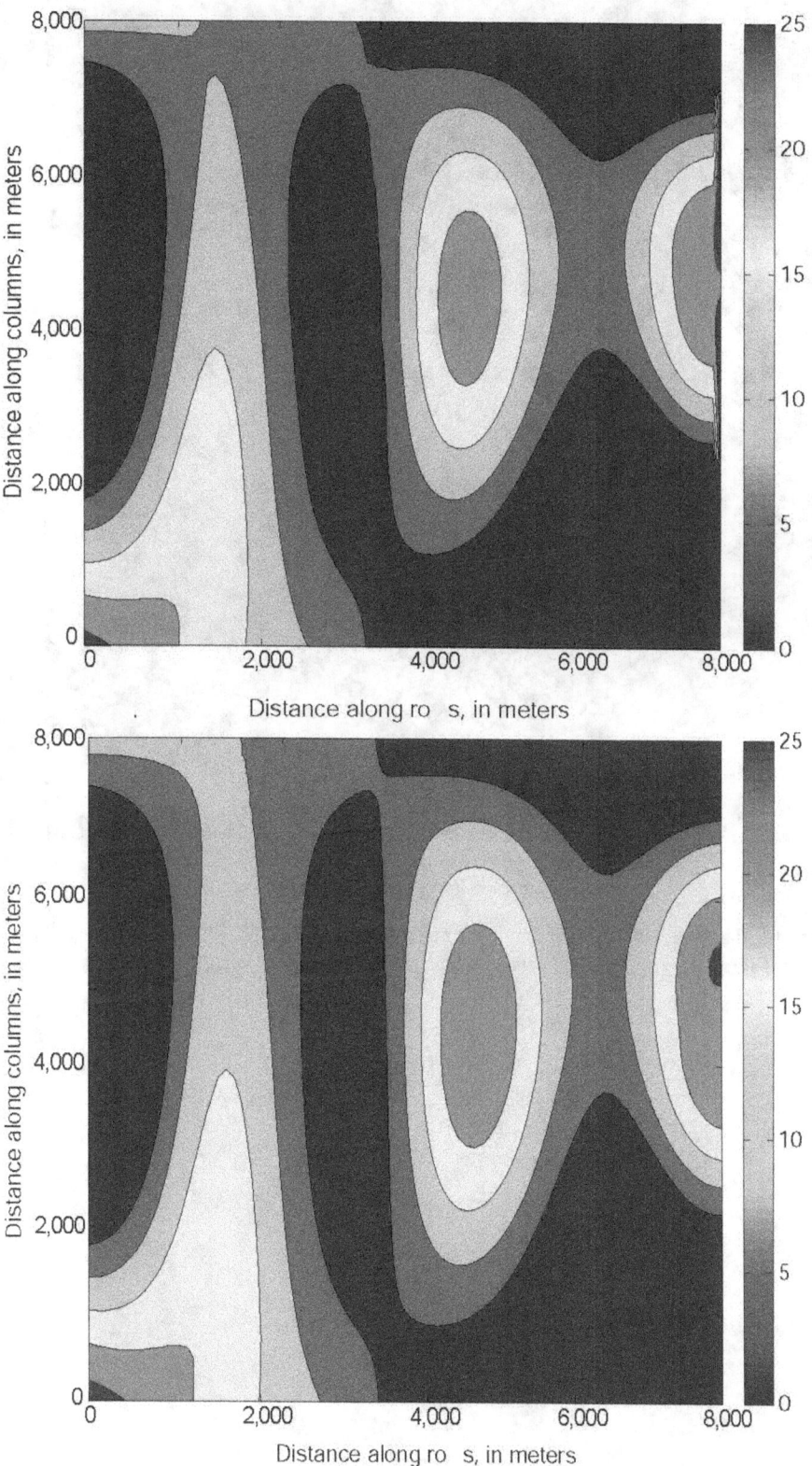

Figure 13.   Distribution of saturated thickness for problem 3 with higher recharge values (fig. 10). (*A*) MODFLOW-2005, and (*B*) MODFLOW-NWT. Units used for saturated thickness are meters above cell bottom.

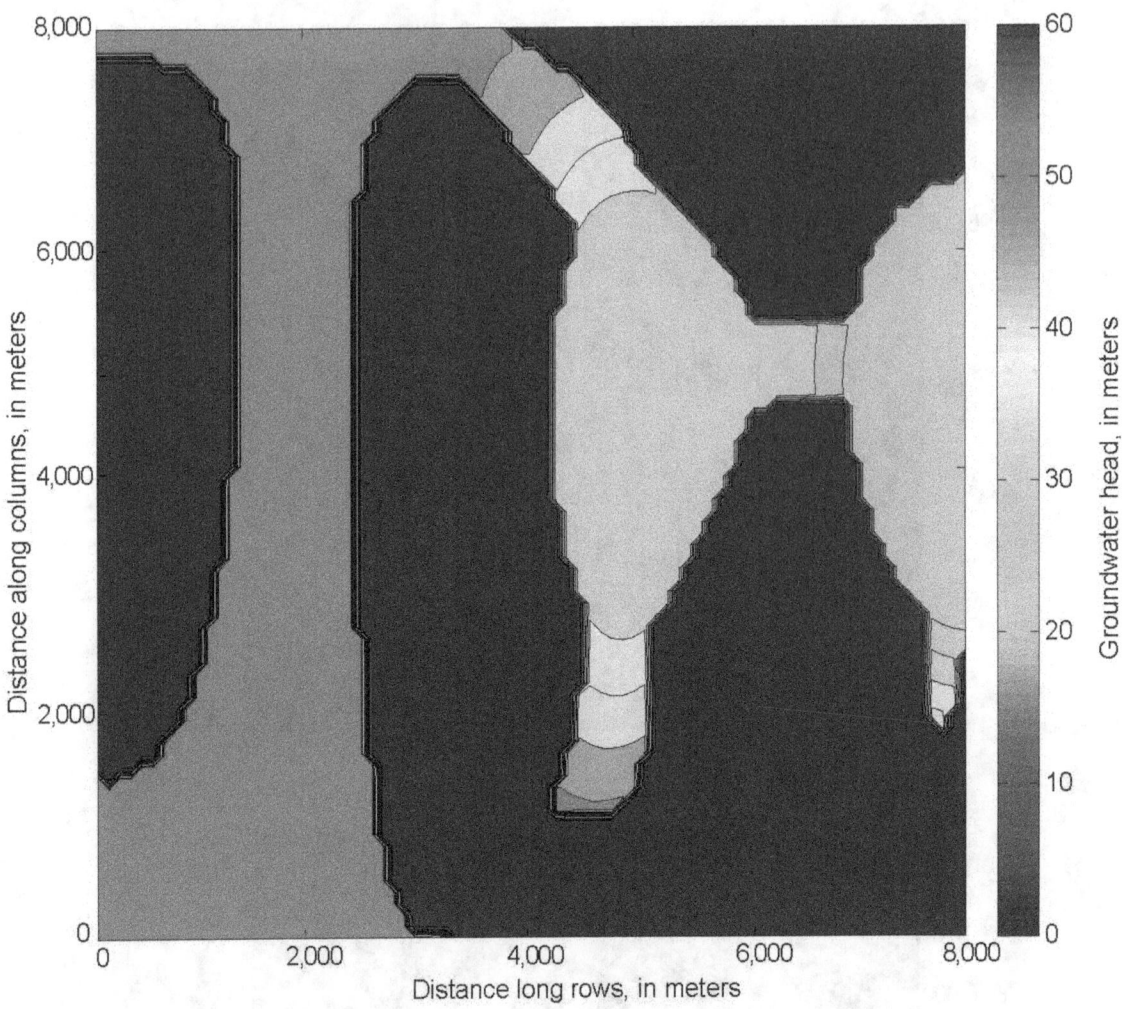

Figure 14.   Distribution of groundwater heads calculated by MODFLOW-NWT for problem 3 with lower recharge values (fig. 10). The dark blue regions indicate cells that have heads less than 2 mm above the cell bottom.

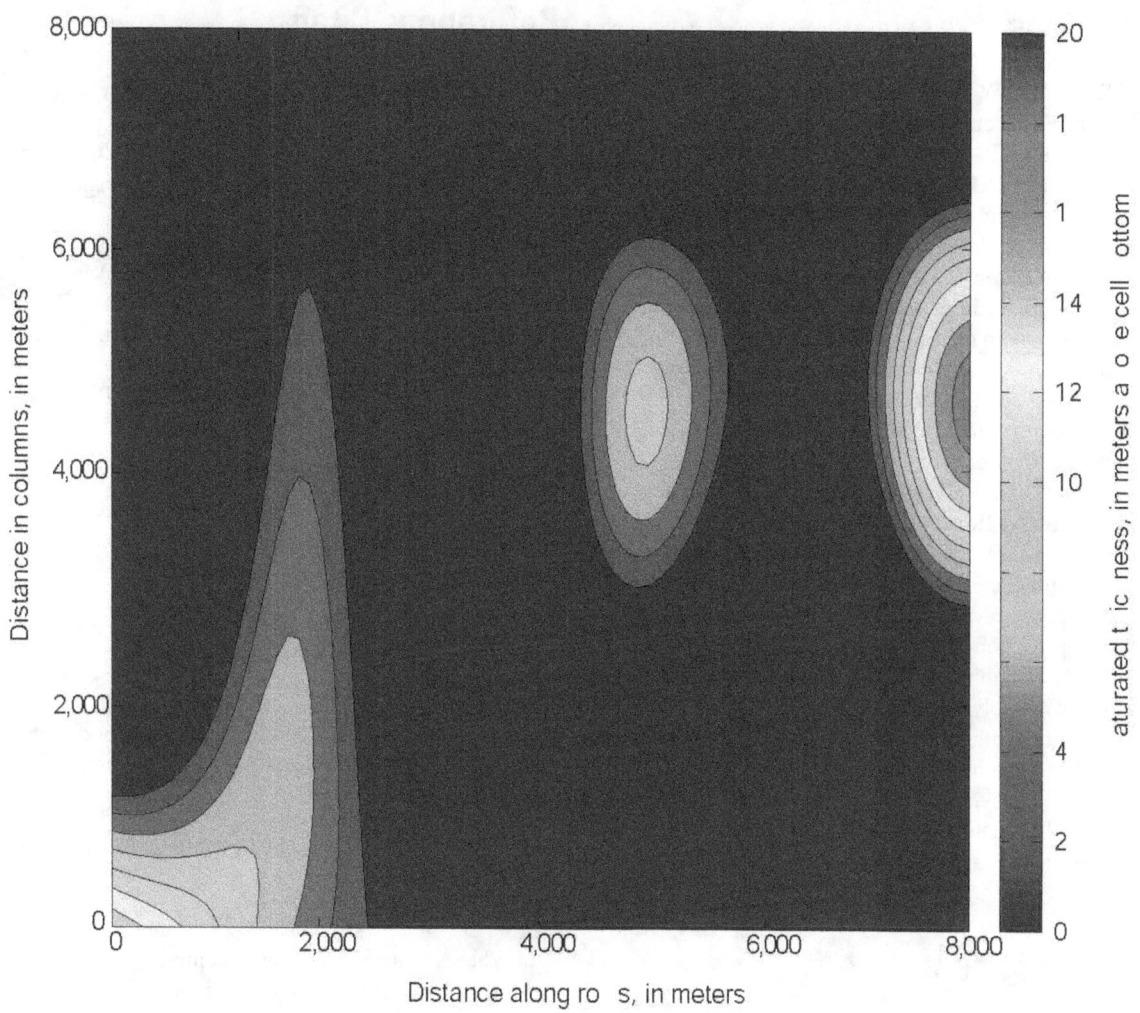

Figure 15. Distribution of saturated thickness calculated by MODFLOW-NWT for problem 3 with lower recharge values (fig. 10).

Conclusions

MODFLOW-NWT simulates three-dimensional groundwater flow through unconfined aquifers while keeping all model cells active during the simulation. Although alternative MODFLOW-2005 approaches are able to solve three-dimensional unconfined problems, this is done by making cells active and inactive during the simulation. Conversion of cells between active and inactive creates discontinuities in calculated flow and can result in convergence failure in many situations. Keeping all cells active within a simulation and using upstream weighting to calculate intercell conductance removes the most common cause of convergence failure, while realistically simulating dewatered cells. Consequently, MODFLOW-NWT will expand the applicability of MODFLOW to a wider variety of applied groundwater-flow problems.

Because MODFLOW-NWT generates asymmetric matrices, additional computations are required. Unlike other MODFLOW-2005 solvers that need to store only one-half of the off-diagonal non-zero elements within the global matrix because of symmetry, all non-zero elements must be stored in the NWT solver. Consequently, MODFLOW-NWT may be slower than other MODFLOW-2005 solvers that rely on the Picard method for some problems. However, the added computations required to solve asymmetric matrices are warranted if MODFLOW-NWT provides solutions for problems that fail to converge using the Picard method.

Acknowledgments

Development of MODFLOW-NWT was supported by the USGS Groundwater Resources Program. Additionally, support for the second author was provided by AMEC Geomatrix Inc. The second author would also like to thank Anthony (Tony) Daus for his support of this work. The authors thank Arlen W. Harbaugh, USGS, emeritus, for his help in the design of MODFLOW-NWT and Eric M. Labolle, University of California, Davis, for his help in providing test models. The authors thank Scott Painter, Southwest Research Institute, Texas, for sharing his ideas on using the Newton method. The authors also thank Thomas E. Reilly, USGS, emeritus, for his help comparing the new schemes with analytical solutions for groundwater flow. The authors also thank Kenneth L. Kipp, USGS, for his help implementing the GMRES Solver into MODFLOW-NWT. The authors also appreciate the thoughtful reviews provided by Steffen Mehl, Professor, California State University, Chico; Ned Banta, USGS; Keith Halford, USGS; Paul Barlow, USGS; Chris Langevin, USGS; Daniel Feinstein, USGS; and John Rupp, Slumberger, Santiago, Chile.

References Cited

Anderman, E.R., and Hill, M.C., 2000, MODFLOW-2000, the U.S. Geological Survey Modular Groundwater Model-Documentation of the Hydrogeologic-Unit Flow (HUF) Package: U.S. Geological Survey Open-File Report 2000-342, 89 p.

Barrett, Richard, Berry, Michael, Chan, T.F., Demmel, James, Donato, June, Dongarra, Jack, Eijkhout, Victor, Pozo, Roldan, Romine, Charles, and van der Vorst, Henk, 1994, Templates for the solution of linear systems—Building blocks for iterative methods: Philadelphia, Penn., Society for Industrial and Applied Mathematics, 112 p.

Celia, M.A., Bouloutas, E.T., Zarba, R.L., 1990, A general mass-conservative numerical solution for the unsaturated flow equation: Water Resources Research, v. 26, no. 7, p. 1483-1496.

Cooley, R.L., 1983, Some new procedures for numerical solution of variably saturated flow problems: Water Resources Research, v. 19, no. 5, p. 1271-1285.

Doherty, John, 2001, Improved calculations for dewatered cells in MODFLOW: Ground Water, v. 39, no. 6, p. 863-869.

Fetter, C.W., 1994, Applied hydrogeology: Columbus, Ohio, Merrill Publishing Company, 592 p.

Goode, D.J., and Appel, C.A., 1992, Finite-difference interblock transmissivity for unconfined aquifers and for aquifers having smoothly varying transmissivity: U.S. Geological Survey Water-Resources Investigation Report 92-4124, 79 p.

Greenbaum, Anne, 1997, Iterative methods for solving linear systems: Philadelphia, Penn., Society for Industrial and Applied Mathematics, 220 p.

Halford, K.J., and Hanson, R.T., 2002, User guide for the drawdown-limited, multi-node well (MNW) package for the U.S. Geological Survey's modular three-dimensional finite difference groundwater flow model, versions MODFLOW-96 and MODFLOW-2000: U.S. Geological Survey Open-File Report 2002-293, 33 p.

Harbaugh, A.W., 2005, MODFLOW-2005, the U.S. Geological Survey modular groundwater model—the Groundwater Flow Process: U.S. Geological Survey Techniques and Methods 6-A16, variously paginated.

Harbaugh, A.W., Banta, E.R., Hill, M.C., and McDonald, M.G., 2000, MODFLOW-2000, The U.S. Geological Survey modular ground-water model—user guide to modularization concepts and the ground-water flow process: U.S. Geological Survey Open-File Report 00–92, 130 p.

Huyakorn, P.S., Springer, E.P., Guvanasen, V., and Wadsworth, T.D., 1986, A three-dimensional finite-element model for simulating water flow in variably saturated porous-media: Water Resources Research, v. 22, no. 13, p. 1790-1808.

HydroGeoLogic, 1996, MODFLOW-SURFACT 99 user manual, Herndon, Virginia: HydroGeoLogic, ISBN 0521880688, 9780521880688, 1235 p.

Jacob, C.E., 1950, Flow of groundwater, chapter 5, *in* Rouse, H., Engineering Hydraulics: John Wiley, New York, p. 321-386.

Jacobs, R.A., 1988, Increased rates of convergence through learning rate adaptation: Neural Networks, v. 1, p. 295-307.

Keating, Elizabeth, and Zyvolosk, George, 2009, A stable and efficient numerical algorithm for unconfined zquifer analysis: Ground Water, v. 47, no. 4, p. 569-579.

Kelley, C.T., 1995, Iterative methods for linear and nonlinear equations: Philadelphia, Penn., Society for Industrial and Applied Mathematics, 166 p.

Kipp, K.L., Jr., Hsieh, P.A., and Charlton, S.R., 2008, Guide to the revised groundwater flow and heat transport simulator: HYDROTHERM — Version 3: U.S. Geological Survey Techniques and Methods 6–A25, 160 p.

Knoll, D.A., and Keyes, D.E., 2004, Jacobian-free Newton–Krylov methods: a survey of approaches and applications: Journal of Computational Physics, 193, issue 2, p. 357-397.

Konikow, L.F., Hornberger, G.Z., Halford, K.J., and Hanson, R.T., 2009, Revised multi-node well (MNW2) package for MODFLOW ground-water flow model: U.S. Geological Survey Techniques and Methods 6–A30, 67 p.

Kuiper, L.K., 1987, A comparison of iterative methods as applied to the solution of the nonlinear three-dimensional groundwater flow equation: SIAM Journal on Scientific and Statistical Computing, v. 8, no. 4, p. 521–528.

Maxwell, R.M., and Miller, N.L., 2005, Development of a coupled land surface and groundwater model: Journal of Hydrometeorology., v. 6, issue 3, p. 233–247.

McDonald, M.G., and Harbaugh, A.W., 1988, A modular three-dimensional finite-difference groundwater flow model: U.S. Geological Survey Techniques of Water-Resources Investigation, Book 6, Chapter A-1, 586 p.

McDonald, M.G., Harbaugh, A.W., Orr, B.R., and Ackerman, D.J., 1991, A method of converting no-flow cells to variable-head cells for the U.S. Geological Survey modular finite-difference groundwater flow model: U.S. Geological Survey Open-File Report 91-536, 99 p.

Mehl, Steffen, 2006, Use of Picard and Newton iteration for solving nonlinear groundwater flow equations: Ground Water, v. 44, no. 4, p. 583-594.

Merritt, M.L., and Konikow, L.F., 2000, Documentation of a computer program to simulate lake-aquifer interaction using the MODFLOW groundwater flow model and the MOC3D solute-transport model: U.S. Geological Survey Water-Resources Investigations Report 00-4167, 146 p.

Niswonger, R.G., and Prudic, D.E., 2005, Documentation of the Streamflow-Routing (SFR2) Package to include unsaturated flow beneath streams, a modification to SFR1: U.S. Geological Survey Techniques and Methods, Book 6, Chapter A13, 62 p.

Niswonger, R.G., Prudic, D.E., and Regan, R.S., 2006, Documentation of the Unsaturated-Zone Flow (UZF1) Package for modeling unsaturated flow between the land surface and the water table with MODFLOW-2005: U.S. Geological Survey Techniques and Methods, Book 6, Chapter A19, 62 p.

Painter, Scott, Başağaoğlu, Hakan, and Liu, A.G., 2008, Robust representation of dry cells in single-layer MODFLOW models: Ground Water, v. 46, no. 6, p. 873-881.

Panday, Sorab, and Huyakorn, P.S., 2004, A fully coupled physically-based spatially-distributed model for evaluating surface/subsurface flow: Advances in Water Resources 27, no. 4, p. 361-382.

Panday, Sorab, and Huyakorn, P.S., 2008, MODFLOW SURFACT: A state-of-the-art use of vadose zone flow and transport equations and numerical techniques for environmental evaluations: Vadose Zone Journal, v. 7, no. 2, p. 610–631.

Patel, V.A., 1994, Numerical analysis: Orlando, Florida, Harcourt Brace, 652 p.

Press, W.H., 2007, Numerical recipes: The art of scientific computing: Cambridge University Press, 1256 p.

Pruess, Karsten., Oldenburg, C.M., and Moridis, G.J., 1999, TOUGH2 user's guide, version 2.0: Report No. LBNL 43134, Ernest Orlando Lawrence Berkeley National Laboratory, Calif., USA.

Saad, Yousef, 1990, Sparskit—A basic tool kit for sparse matrix computations: Technical Report RIACS–90–20, Moffet Field, Calif., Research Institute for Advanced Computer Science, NASA Ames Research Center, Software available at: http://www.ca.umn.edu/~saad/software.

Saad, Yousef, 2003, Iterative methods for sparse linear systems (2d ed.): Philadelphia, Penn., Society for Industrial and Applied Mathematics, 528 p.

Smith, Murray, 1993, Neural networks for statistical modeling: Van Nostrand Reinhold, New York, 235 p.

Todd, D.K., and Mays, L.W., 2005, Groundwater hydrology (Third Edition): New York, John Wiley and Sons, 636 p.

van der Vorst, H.A., 1992, Bi-CGSTAB: A fast and smoothly converging variant of Bi-CG for the solution of nonsymmetric linear systems: SIAM Journal on Scientific and Statistical Computing, v. 13, no. 2, p. 631-644.

Appendix A. Input Instructions and List File for MODFLOW-NWT

Most of the input-file structures that are needed for a MODFLOW-NWT simulation are unchanged from those documented for MODFLOW-2005. Therefore, only the input instructions for those packages that are new or must be modified for use with MODFLOW-NWT are described in this Appendix.

MODFLOW-NWT Name File

The Name File contains the names of most of the input and output files used in a model simulation and controls whether or not parts of the model program are active. The MODFLOW-NWT Name File has the same function and input format as the MODFLOW-2005 Name File. Table 1 lists the file types that are supported by MODFLOW-NWT. The "NWT" and "UPW" input-file types must be included in the MODFLOW-NWT Name File if the Newton method is used. No other internal flow package file types (that is, BCF6, LPF, or HUF) other than "UPW" can be included in the MODFLOW-NWT Name File if Newton linearization is used. Additionally, none of the MODFLOW-2005 solver input file types can be included in the MODFLOW-NWT Name File (that is, SIP, PCG, or DE4) if Newton linearization is used. However, standard flow packages (BCF, LPF, and HUF Packages) and solver packages (SIP, PCG, and DE4 Packages) can be used if the groundwater-flow equation is solved using MODFLOW-2005 solution method; for this case, MODFLOW-NWT is identical to MODFLOW-2005. Also, as is the case for MODFLOW-2005, "OPEN/CLOSE" file types, which are described in the Input Instructions for Array Reading Utility Subroutines on pages 8-57 through 8-60 in Harbaugh (2005) are not included in the Name File.

UPW Package Input

Cell property data are read from the file that is type "UPW" in the Name File. Free format is used for reading all values. Input data types and formats described below are nearly identical to those for the LPF Package and described on pages 8-28 through 8-31 of the MODFLOW-2005 manual (Harbaugh, 2005). However, the input variable LAYWET should be set to zero for all layers because all layers that are specified as convertible (that is, LAYTYP>0) are assumed to be wettable in the UPW Package. The model will stop and print an error statement to the main Listing file if LAYWET is non-zero for any layers. Options are not allowed in the UPW Package, as they are in the LPF Package, and the variable IPHDRY is new to the UPW Package.

For Each Simlation

0. [#Text]

Item 0 is optional—"#" must be in column 1. Item 0 can be repeated multiple times.

1. IUPWCB HDRY NPUPW IPHDRY
2. LAYTYP(NLAY)
3. LAYAVG(NLAY)
4. CHANI(NLAY)
5. LAYVKA(NLAY)
6. LAYWET(NLAY)

(Item 6 should always be set to zero in the UPW Package because all layers with LAYTYP(NLAY) >0 are assumed to be wettable)

7. [PARNAM PARTYP Parval NCLU]
8. [Layer Mltarr Zonarr IZ]

Each repetition of Item 8 is called a parameter cluster. Repeat Item 8 NCLU times.

Repeat Items 7-8 for each parameter to be defined (that is, NPUPW times).

A subset of the following two-dimensional variables is used to describe each layer. All variables that apply to layer 1 are read first, followed by layer 2, followed by layer 3, and so forth. A variable not required due to simulation options (for example, Ss and Sy for a completely steady-state simulation) must be omitted from the input file.

These variables are either read by the array-reading utility subroutine, U2DREL, or they are defined through parameters. If a variable is defined through parameters, then the variable itself is not read; however, a single line containing a print code is read in place of the control line. The print code determines the format for printing the values of the variable as defined by parameters. The print codes are the same as those used in a control line. If any parameters of a given type are used, parameters must be used to define the corresponding variable for all layers in the model.

9. HK(NCOL,NROW)   If any HK parameters are included, read only a print code.

10. [HANI(NCOL,NROW)]   Include item 10 only if CHANI is less than or equal to 0. If any HANI parameters are included, read only a print code.

11. VKA(NCOL,NROW)   If any VK or VANI parameters are included, read only a print code.

12. [Ss(NCOL,NROW)]   Include item 12 only if at least one stress period is transient. If there are any SS parameters, read only a print code.

13. [Sy(NCOL,NROW)]   Include item 13 only if at least one stress period is transient and LAYTYP >0. If any SY parameters are included, read only a print code.

14. [VKCB(NCOL,NROW)]   Include item 14 only if LAYCBD (in the Discretization File) is not 0. If any VKCB parameters are included, read only a print code.

Explanation of Variables Read by the UPW Package:

Text   is a character variable (199 characters) that starts in column 2. Any characters can be included in Text. The "#" character must be in column 1. Lines beginning with # are restricted to the first lines of the file. Text is written to the Listing File.

IUPWCB   is a flag and a unit number.

If IUPWCB > 0, cell-by-cell flow terms will be written to this unit number when "SAVE BUDGET" or a non-zero value for ICBCFL is specified in Output Control. The terms that are saved are storage, constant-head flow, and flow between adjacent cells.

If IUPWCB = 0, cell-by-cell flow terms will not be written.

If IUPWCB < 0, cell-by-cell flow for constant-head cells will be written in the listing file when "SAVE BUDGET" or a non-zero value for ICBCFL is specified in Output Control. Cell-by-cell flow to storage and between adjacent cells will not be written to any file.

HDRY   is the head that is assigned to cells that are converted to dry during a simulation. Although this value plays no role in the model calculations, HDRY values are useful as indicators when looking at the resulting heads that are output from the model. HDRY is thus similar to HNOFLO in the Basic Package, which is the value assigned to cells that are no-flow cells at the start of a model simulation.

NPUPW   is the number of UPW parameters.

IPHDRY   is a flag that indicates whether groundwater head will be set to HDRY when the groundwater head is less than 1×10^{-4} above the cell bottom (units defined by LENUNI). If IPHDRY=0, then head will not be set to HDRY. If IPHDRY>0, then head will be set to HDRY.

LAYTYP   contains a flag for each layer that specifies the layer type.
                    0 – confined
                    >0 – convertible
                    <0 – confined

LAYAVG   contains a flag for each layer that defines the method of calculating interblock transmissivity.
                    0—harmonic mean
                    1—logarithmic mean
                    2—arithmetic mean of saturated thickness and logarithmic-mean hydraulic conductivity.

CHANI   contains a value for each layer that is a flag or the horizontal anisotropy. If CHANI is less than or equal to 0, then variable HANI defines horizontal anisotropy. If CHANI is greater than 0, then CHANI is the horizontal anisotropy for the entire layer, and HANI is not read. If any HANI parameters are used, CHANI for all layers must be less than or equal to 0.

LAYVKA   contains a flag for each layer that indicates whether variable VKA is vertical hydraulic conductivity or the ratio of horizontal to vertical hydraulic conductivity.
0—indicates VKA is vertical hydraulic conductivity
not 0—indicates VKA is the ratio of horizontal to vertical hydraulic conductivity, where the horizontal hydraulic conductivity is specified as HK in item 9.

LAYWET   contains a flag for each layer that indicates whether wetting is active. LAYWET should always be zero for the UPW Package because all cells initially active are wettable.

PARNAM   is the name of a parameter to be defined. This name can consist of 1 to 10 characters and is not case sensitive. That is, any combination of the same characters with different case will be equivalent.

PARTYP   is the type of parameter to be defined. For the UPW Package, the allowed parameter types are:
HK—defines variable HK, horizontal hydraulic conductivity
HANI—defines variable HANI, horizontal anisotropy
VK—defines variable VKA for layers for which VKA represents vertical hydraulic conductivity (LAYVKA=0)
VANI—defines variable VKA for layers for which VKA represents vertical anisotropy (LAYVKA≠0)
SS—defines variable Ss, the specific storage
SY—defines variable Sy, the specific yield
VKCB—defines variable VKCB, the vertical hydraulic conductivity of a Quasi-3D confining layer.

Parval   is the parameter value. This parameter value may be overridden by a value in the Parameter Value File.

NCLU   is the number of clusters required to define the parameter. Each repetition of Item 8 is a cluster (variables Layer, Mltarr, Zonarr, and IZ). Each layer that is associated with a parameter usually has only one cluster. For example, parameters which apply to cells in a single layer generally will be defined by just one cluster. However, having more than one cluster for the same layer is acceptable.

Layer   is the layer number to which a cluster definition applies.

Mltarr   is the name of the multiplier array to be used to define variable values that are associated with a parameter. The name "NONE" means that there is no multiplier array, and the variable values will be set equal to Parval.

Zonarr   is the name of the zone array to be used to define the cells that are associated with a parameter. The name "ALL" means that there is no zone array, and all cells in the specified layer are part of the parameter.

IZ   is up to 10 zone numbers (separated by spaces) that define the cells that are associated with a parameter. These values are not used if ZONARR is specified as "ALL". Values can be positive or negative, but 0 is not allowed. The end of the line, a zero value, or a non-numeric entry terminates the list of values.

HK   is the hydraulic conductivity along rows. HK is multiplied by horizontal anisotropy (see CHANI and HANI) to obtain hydraulic conductivity along columns.

HANI   is the ratio of hydraulic conductivity along columns to hydraulic conductivity along rows, where HK of item 9 specifies the hydraulic conductivity along rows. Thus, the hydraulic conductivity along columns is the product of the values in HK and HANI. Read only if CHANI ≤ 0.

VKA   is either vertical hydraulic conductivity or the ratio of horizontal to vertical hydraulic conductivity depending on the value of LAYVKA. If LAYVKA is 0, VKA is vertical hydraulic conductivity. If LAYVKA is not 0, VKA is the ratio of horizontal to vertical hydraulic conductivity. In this case, HK is divided by VKA to obtain vertical hydraulic conductivity, and values of VKA typically are greater than or equal to 1.0.

Ss   is specific storage. Read only for a transient simulation (at least one transient stress period).

Sy   is specific yield. Read only for a transient simulation (at least one transient stress period) and if the layer is
              convertible (LAYTYP >0).

VKCB   is the vertical hydraulic conductivity of a quasi-three-dimensional confining bed below a layer. Read only if
              there is a confining bed. Because the bottom layer cannot have a confining bed, VKCB cannot be specified for
              the bottom layer.

NWT Input File

Input to the Newton (NWT) Solver Package is read from the file that is type "NWT" in the Name File. All numeric
variables are free format if the option "FREE" is specified in the Basic Package input file; otherwise, all variables have
10-character fields. Values in brackets are specified only for some input options. If the OPTIONS keyword is specified as
"SIMPLE," "MODERATE," or "COMPLEX," then all solver input variables following OPTIONS in item 1 are not required
to be specified and are set to default values. If OPTIONS is specified as "SPECIFIED," then all input variables are required;
however, residual-control variables are only required if BACKFLAG is set to one. The residual-control option can only be used if
OPTIONS is specified as "SPECIFIED." It is recommended that users refer to table 2 and the discussion on "Example Input for
the NWT Input File" for guidance on values that might be identified for each input variable.

For Each Simulation

0. [#Text]

Item 0 is optional—"#" must be in column 1. Item 0 can be repeated multiple times.

1. HEADTOL FLUXTOL MAXITEROUT THICKFACT LINMETH IPRNWT IBOTAV OPTIONS
   [DBDTHETA] [DBDKAPPA] [DBDGAMMA] [MOMFACT] [BACKFLAG] [MAXBACKITER]
   [BACKTOL] [BACKREDUCE]

If LINMETH = 1 and OPTION is set to "SPECIFIED" read the following line:

2a. [MAXITINNER] [ILUMETHOD][LEVFILL][STOPTOL][MSDR]

If LINMETH = 2 and OPTION is set to "SPECIFIED" read the following line:

2b. [IACL][NORDER][LEVEL][NORTH][IREDSYS][RRCTOLS][IDROPTOL][EPSRN]
    [HCLOSEXMD][MXITERXMD]

Explanation of Variables Read by the NWT Solver

Text   is a character variable (199 characters) that starts in column 2. Any characters can be included in Text. The
              "#"character must be in column 1. Lines beginning with # are restricted to the first lines of the file. Text is
              written to the Listing File.

HEADTOL   (units of length)—is the maximum head change between outer iterations for solution of the nonlinear
              problem (real).

FLUXTOL   (units of length cubed per time)—is the maximum root-mean-squared flux difference between outer iterations
              for solution of the nonlinear problem (real).

MAXITEROUT   is the maximum number of iterations to be allowed for solution of the outer (nonlinear) problem (integer).

THICKFACT   is the portion of the cell thickness (length) used for smoothly adjusting storage and conductance coefficients
              to zero (symbol Ω in equation 9; real).

LINMETH   is a flag that determines which matrix solver will be used. A value of 1 indicates GMRES will be used and a
              value of 2 indicates χMD will be used (integer).

IPRNWT   is a flag that indicates whether additional information about solver convergence will be printed to the main listing file (integer).

IBOTAV   is a flag that indicates whether corrections will be made to groundwater head relative to the cell-bottom altitude if the cell is surrounded by dewatered cells (integer). A value of 1 indicates that a correction will be made and a value of 0 indicates no correction will be made. This input variable is problem specific and both options (IBOTAV=0 or 1) should be tested.

OPTIONS   are keywords that activate options:

SPECIFIED indicates that the optional solver input values listed for items 1 and 2 will be specified in the NWT input file by the user.

SIMPLE indicates that default solver input values will be defined that work well for nearly linear models. This would be used for models that do not include nonlinear stress packages, and models that are either confined or consist of a single unconfined layer that is thick enough to contain the water table within a single layer. (See table 2 for the solver input values that will be used for this option.)

MODERATE indicates that default solver input values will be defined that work well for moderately nonlinear models. This would be used for models that include nonlinear stress packages, and models that consist of one or more unconfined layers. The "MODERATE" option should be used when the "SIMPLE" option does not result in successful convergence. (See table 2 for the solver input values that will be used for this option.)

COMPLEX indicates that default solver input values will be defined that work well for highly nonlinear models. This would be used for models that include nonlinear stress packages, and models that consist of one or more unconfined layers representing complex geology and sw/gw interaction. The "COMPLEX" option should be used when the "MODERATE" option does not result in successful convergence. (See table 2 for the solver input values that will be used for this option.)

Read the following values if OPTIONS = **"SPECIFIED."**

DBDTHETA   is a coefficient used to reduce the weight applied to the head change between nonlinear iterations (symbol θ in equation 21). DBDTHETA is used to control oscillations in head. Values range between 0.0 and 1.0, and larger values increase the weight (decrease under-relaxation) applied to the head change (real).

DBDKAPPA   is a coefficient used to increase the weight applied to the head change between nonlinear iterations (symbol κ in equation 22). DBDKAPPA is used to control oscillations in head. Values range between 0.0 and 1.0, and larger values increase the weight applied to the head change (real).

DBDGAMMA   is a factor (symbol γ in equation 19) used to weight the head change for iterations $n-1$ and n. Values range between 0.0 and 1.0, and greater values apply more weight to the head change calculated during iteration n (real).

MOMFACT   is the momentum coefficient m of equation 20 and ranges between 0.0 and 1.0. Greater values apply more weight to the head change for iteration n (real).

BACKFLAG   is a flag used to specify whether residual control will be used. A value of 1 indicates that residual control is active and a value of 0 indicates residual control is inactive (integer).

MAXBACKITER is the maximum number of reductions (backtracks) in the head change between nonlinear iterations (integer). A value between 10 and 50 works well.

BACKTOL   is the proportional decrease in the root-mean-squared error of the groundwater-flow equation used to determine if residual control is required at the end of a nonlinear iteration, as applied in equation 23 (real).

BACKREDUCE is a reduction factor (symbol B_r in equation 23) used for residual control that reduces the head change between nonlinear iterations (real). Values should be between 0.0 and 1.0, where smaller values result in smaller head-change values.

Read the following values if LINMETH = 1 **and** OPTIONS = **"SPECIFIED."**

MAXITINNER   is the maximum number of iterations for the linear solution (integer).

ILUMETHOD   is the index for selection of the method for incomplete factorization (ILU) used as a preconditioner. See Kipp and others (2008) for further details (integer).

      ILUMETHOD=1—ILU with drop tolerance and fill limit. Fill-in terms less than drop tolerance times the diagonal are discarded. The number of fill-in terms in each row of L and U is limited to the fill limit. The fill-limit largest elements are kept in the L and U factors.

      ILUMETHOD=2 — ILU(k), Order k incomplete LU factorization. Fill-in terms of higher order than k in the factorization are discarded.

LEVFILL   is the fill limit for ILUMETHOD = 1 and is the level of fill for ILUMETHOD = 2. Recommended values: 5-10 for method 1, 0-2 for method 2. See Kipp and others (2008) for further details (integer).

STOPTOL   is the tolerance for convergence of the linear solver. This is the residual of the linear equations scaled by the norm of the root mean squared error. Usually 10^{-8} to 10^{-12} works well. See Kipp and others (2008) for further details (integer).

MSDR   is the number of iterations between restarts of the GMRES Solver. See Kipp and others (2008) for further details (integer).

Read the following values if LINMETH = 2 **and** OPTIONS="**SPECIFIED.**"

IACL   is a flag for the acceleration method: 0 = conjugate gradient; 1 = ORTHOMIN; 2 = Bi-CGSTAB (integer).

NORDER   is a flag for the scheme of ordering the unknowns: 0= original ordering; 1= RCM ordering; 2= Minimum Degree ordering (integer).

LEVEL   is the level of fill for incomplete LU factorization (integer).

NORTH   is the number of orthogonalization for the ORTHOMIN acceleration scheme. A number between 4 and 10 is appropriate. Small values require less storage but more iterations may be required. This number should equal 2 for the other acceleration methods (integer).

IREDSYS   is a flag for reduced system preconditioning: =1 apply reduced system preconditioning; = 0 do not apply reduced system preconditioning (integer).

RRCTOLS   is the residual reduction-convergence criteria (real).

IDROPTOL   is a flag for using drop tolerance in the preconditioning (integer).

EPSRN   is the drop tolerance for preconditioning (real).

HCLOSEXMD   is the head closure criteria for inner (linear) iterations (real).

MXITERXMD   is the maximum number of iterations for the linear solution.

WEL Package Input File

The WEL Package input file remains unchanged, unless the user wants to specify the fraction of the cell thickness (input variable PHIRAMP) that is used to calculate the pumping rate reduction interval (variable Φ in the variable definitions for equations 24 and 25). If PHIRAMP is not specified, then a default value of 0.2 is used. The value PHIRAMP is read when the keyword SPECIFY is specified as item 1b of the WEL Package input file. The value PHIRAMP is specified following SPECIFY in item 2b. The modification made to the WEL Package follows Item 2 of the input file (see p. 8-40 of Harbaugh, 2005):

2a. Data MXACTW IWELCB [Option]

2b. Data [SPECIFY] [PHIRAMP]

XMACTW and IWELCB definitions are unchanged (see p. 8-41 of Harbaugh, 2005).

SPECIFY   is a keyword option used to specify a value for PHIRAMP. If SPECIFY is not specified then a default value of 0.05 is used.

PHIRAMP   is the fraction of the cell thickness used as an interval for smoothly adjusting negative pumping rates to 0 for dry cells. Negative pumping rates are adjusted to 0 or a smaller negative value when the head in the cell is equal to or less than the calculated interval above the cell bottom (see equations 24 and 25 on page 14).

MODFLOW-NWT Listing File

The MODFLOW-NWT Listing File may include printouts of any selected input read by the model—groundwater head, drawdown, and model budget reports; model progress, including time step and stress period information; and solver information. Additional information related to input and output from stress packages also may be printed to the Listing File.

The MODFLOW-NWT Listing File is nearly identical to the MODFLOW-2005 Listing File, except for two differences. The first difference is the output for the solver describing inner and outer iterations. Figure A1 shows a sample printout of

the solver information that is printed to the Listing File if the NWT input variable IPRNWT is set to 1. The value printed under the column titled "Maximum-Head-Change" must be less than the NWT input variable HEADTOL, and the value in the column titled "RMS" must be less than the NWT input variable FLUXTOL for convergence. The value printed directly below "Maximum-Flux-Residual" is the volumetric flux residual for the cell with the largest absolute residual value. The second difference in the MODFLOW-NWT Listing file is the printing of the actual pumping rate for unconfined cells that cannot produce enough water to maintain the specified pumping rate (fig. A2).

```
SOLVING FOR HEAD

                                        Max.-Head-Change               Max.-Flux-Residual
Residual-Control  Outer-Iter.  Inner-Iter.  Column Row Layer  Maximum-Head-Change  Column Row Layer  Maximum-Flux-Residual        RMS
      0                1           185          1    1    1     0.7281980658E+08      80   49    1      0.3823164204E+03     0.8425127560E+03
      0                2            42          1    1    1    -0.7281972894E+08       1    1    1     -0.1035564882E+11     0.1145324689E+11
      0                3            30         58   42    1    -0.3519553756E+02      54   34    1     -0.5992673473E+04     0.4227542585E+05
      0                4            27         74   63    1     0.1690224300E+02      76   61    1     -0.1546020073E+04     0.1106078571E+05
      0                5            28         74   66    1     0.1002396496E+02      76   61    1     -0.5336831171E+01     0.2878673157E+04
      0                6            26         73   68    1     0.4633234260E+01      76   61    1     -0.1579939918E+03     0.6710117840E+03
      0                7            18         73   70    1     0.1578516236E+01      76   61    1     -0.3709063306E+02     0.1182220337E+03
      0                8            18          9   56    1     0.9111083414E+00      76   61    1     -0.7443316869E+01     0.2012351434E+02
      0                9             6          5   28    1    -0.2721178690E+00       9   54    1     -0.1533958512E+01     0.4527352406E+01
      0               10             4          5   28    1    -0.1173265997E+00       9   54    1     -0.5314328707E+00     0.1260880644E+01
                                        Max.-Head-Change               Max.-Flux-Residual
Residual-Control  Outer-Iter.  Inner-Iter.  Column Row Layer  Maximum-Head-Change  Column Row Layer  Maximum-Flux-Residual        RMS
      0               11             2          5   28    1    -0.4303239817E-01       9   54    1     -0.1707842198E+00     0.3398471721E+00
      0               12             2          5   28    1    -0.1254800568E-01       9   54    1     -0.4528582975E-01     0.8836627161E-01
      0               13             2          5   28    1    -0.3230607774E-02       9   54    1     -0.1139955925E-01     0.2222415532E-01
      0               14             4          5   28    1    -0.8300484139E-03       9   54    1     -0.2822213133E-02     0.5525538676E-02

       -----------------------------------------------
       NWT REQUIRED         15 OUTER ITERATIONS
       AND A TOTAL OF      394 INNER ITERATIONS.
       -----------------------------------------------
```

Figure A1. Linear (inner) and nonlinear (outer) iteration information printed to the MODFLOW-NWT Listing File.

```
TO AVOID PUMPING WATER FROM A DRY CELL
THE PUMPING RATE WAS REDUCED FOR CELL(IC,IR,IL)    175    1    4
THE SPECIFIED RATE IS    -1.05500    AND THE REDUCED RATE IS    -1.04479
THE HEAD IS    9.617884E+02 AND THE CELL BOTTOM IS    9.523600E+02
TO AVOID PUMPING WATER FROM A DRY CELL
THE PUMPING RATE WAS REDUCED FOR CELL(IC,IR,IL)    176    1    4
THE SPECIFIED RATE IS    -1.05500    AND THE REDUCED RATE IS    -1.04668
THE HEAD IS    9.618456E+02 AND THE CELL BOTTOM IS    9.523600E+02
TO AVOID PUMPING WATER FROM A DRY CELL
THE PUMPING RATE WAS REDUCED FOR CELL(IC,IR,IL)    177    1    4
THE SPECIFIED RATE IS    -1.05500    AND THE REDUCED RATE IS    -1.04781
THE HEAD IS    9.618830E+02 AND THE CELL BOTTOM IS    9.523600E+02
TO AVOID PUMPING WATER FROM A DRY CELL
THE PUMPING RATE WAS REDUCED FOR CELL(IC,IR,IL)    178    1    4
THE SPECIFIED RATE IS    -1.05500    AND THE REDUCED RATE IS    -1.04825
THE HEAD IS    9.618985E+02 AND THE CELL BOTTOM IS    9.523600E+02
```

Figure A2. Message indicating that the pumping rate was reduced due to near-dry conditions in an unconfined model cell.

Appendix B. Example for Modifying Nonlinear-Stress Packages for MODFLOW-NWT

Nonlinear stress packages can be used with MODFLOW-NWT without modification. However, adding the stress derivative to the Jacobian in MODFLOW-NWT can enhance convergence. The following example presents the steps that are required for adding stress package derivatives to MODFLOW-NWT.

Consider the simple quadratic function that relates flow into a cell as a function of head:

$$Q = f(h) = ah^2 + bh, \tag{b1}$$

where Q is the volumetric flow rate into a cell, h is the groundwater head, and a and b are coefficients. The Newton method is applied as:

$$\frac{\partial}{\partial h} f(h) \Delta h = -f(h), \tag{b2}$$

and

$$\frac{\partial}{\partial h} f(h) = \frac{\partial}{\partial h} (ah^2 + bh) = 2ah + b. \tag{b3}$$

In this case, the nonlinear boundary condition has a known functional relationship to h and the analytical derivative can be determined. However, if this function is not known, then a numerical derivative can be used. Equation b2 can be rewritten as

$$(2ah + b) \Delta h = -(ah^2 + bh), \tag{b4}$$

where $\Delta h = h_k - h_{k-1}$, the subscripts k and $k-1$ designate values of h for nonlinear iterations k and $k-1$, and all other occurrences of h in equation b4 are assumed equal to h_{k-1}. After rearranging terms, the equation is put in the form of the equations that are solved by MODFLOW-NWT

$$(2ah_{k-1} + b) h_k = ah_{k-1}^2, \tag{b5}$$

where h_k is the unknown head, and h_{k-1} is the known head from the previous nonlinear iteration. Now that equation b5 is in the same form as the equations solved by MODFLOW-NWT, these terms can be added to the appropriate arrays within the MODFLOW-NWT source code. First, a Fortran "use" statement will need to be added to the formulate subroutine of the nonlinear stress package to access the MODFLOW-NWT arrays, for example

$$\text{USE GWFNWTMODULE, ONLY : A, IA, Icell.} \tag{b6}$$

The variable A is the one-dimensional array that stores the Jacobian matrix, and IA holds the location in A of the first nonzero element of row i of the Jacobian matrix. To access the correct row location corresponding to the cell to which the nonlinear boundary condition is applied, the three-dimensional pointer array Icell is used

$$i = \text{Icell(icol, irow, ilay)}, \tag{b7}$$

where icol, irow, and ilay are the column, row, and layer corresponding to the cell to which the nonlinear boundary condition is applied. Now that the correct row of the Jacobian matrix has been determined, the term multiplied by h_k in equation b5 can be added to the Jacobian

$$\text{A(IA(i))} = \text{A(IA(i))} + 2ah_{k-1} + b, \tag{b8}$$

and the right-hand-side of equation b5 can be added to the RHS array in MODFLOW-NWT

$$\begin{aligned} &\text{RHS(icol,irow,ilay)} = \\ &\text{RHS(icol,irow,ilay)} + ah_{k-1}^2, \end{aligned} \tag{b9}$$

Appendix C. χMD Users' Guide Version 1.3: An Efficient Sparse Matrix Solver Library

Introduction

χMD sparse matrix solver library is designed to solve a symmetric or asymmetric sparse matrix which often arises in numerical modeling approaches in science and engineering. A matrix should be REAL and have non-zero diagonal elements in every row. χMD sparse matrix solver library is designed to solve general real sparse matrices, regardless of their matrix structures. χMD sparse matrix solver library has an option to use reduced system preconditioning that could result in dramatic reduction of computational cost and storage requirement. The matrix solver package consists of three main sections: the preprocessing part, the preconditioning part, and the acceleration part. The preconditioning scheme permits various levels of incomplete LU factorization and reordering of unknowns. In addition to these, the solver package allows the use of a drop tolerance scheme, which reduces computational cost and storage. Various acceleration schemes are used in the acceleration part.

Brief Outline of Theory

The preconditioned conjugate gradient type method (iterative method) has shown its robustness for obtaining a sparse matrix solution (Axelsson, 1977). This method is composed of preconditioning part and acceleration part. Even if the acceleration methods guarantee that the solution will be obtained within a finite number of iterations, higher quality of preconditioning is necessary to reduce the number of iterations and the computational cost.

Most preconditioned conjugate gradient type methods use incomplete LU factorization (ILU) which is similar to Gaussian elimination but decomposition is terminated at a certain level. With a higher level of factorization, fewer iterations are required. However, cost-per-iteration in higher level factorization is higher than that of lower level factorization because a greater number of new fill-in entries appear in the process of decomposition. Therefore, even if the number of iterations is decreased with a higher factorization, the total computational cost is not necessarily reduced.

To obtain a higher level factorization, that is, higher quality of preconditioning and less expensive computational work per iteration, a drop tolerance scheme has been implemented (Munksgaard, 1980; Zlatev, 1982). In this scheme, small new fill-in entries that tend to appear in higher factorization and have little or no effect on the quality of preconditioning are discarded.

Although the use of a drop-tolerance approach is usually advantageous, it may not be in some cases if the original ordering is poor and the decay of the fill-in entries during decomposition is slow. In these cases, the ILU factorization also is poor. The effect of ordering has been investigated through many studies, for example, Behie and Forsyth (1984); Simon (1988); and Young and others (1989).

The reordering of unknowns was originally implemented in the context of direct methods for solving sparse matrices (Duff and others, 1986). The purpose of reordering is to produce a permutation matrix that has less fill-in than the original matrix during the process of Gaussian elimination. This feature is extremely important with regard to ILU factorization. For example, if a permutation matrix produces less fill-in than the original matrix in the process of factorization, then the incompletely-factorized matrix can "inherit" more information than the original matrix contains. Reordering, therefore, can improve the quality of incomplete factorization. χMD uses either original, Minimum degree (Tinney and Walker, 1967) or Reverse Cuthill-McKee orderings (Cuthill and McKee, 1969; George, 1971).

The acceleration module of χMD uses conjugate gradient (Hestenes and Stiefel, 1952), ORTHOMIN (Vinsome, 1976), and CGSTAB (van der Vorst, 1992) accelerations.

Description of Subroutines

χMD sparse matrix solver library contains main subroutines, which appear in Figure C1 as well as utility subroutines. χMD is designed to solve general REAL sparse matrices, regardless of their matrix structures.

Main Subroutines

xmdprpc
Preprocessor for the solver packages. Once the *ia,ja* data structure is set up, this subroutine is called only once.

xmdprecl
Preconditions the coefficient matrix (Symbolic Factorization). This subroutine includes symbolic factorization, which determines the structure of the factorized matrix and is relatively computationally expensive. This subroutine preconditions the coefficient matrix based on specified level of incomplete LU factorization. The symbolic factorization should be done before the numeric factorization is performed.

xmdnfctr

Preconditions the coefficient matrix (Numeric Factorization). This subroutine performs a relatively inexpensive numeric factorization, which numerically factors the matrix based on the structure generated by the symbolic factorization.

xmdprecd

Preconditions the coefficient matrix (both Symbolic and Numeric Factorizations). This subroutine preconditions the coefficient matrix based on the combination of specified level of incomplete LU factorization and drop tolerance. Following call of xmdnfctr is not required because numerical factorization also is performed in this subroutine.

xmdsolv

Acceleration part to obtain solution. This subroutine uses conjugate gradient, ORTHOMIN, and CGSTAB accelerations. Appropriate precondioner must be called before this subroutine is called.

Utility Subroutines

xmddgscl

Scale a matrix. This subroutine transforms a matrix to an identity matrix, whose diagonal entries are equal to one, through multiplication by a proper diagonal matrix. This module modifies the right-hand-side vector **b** as well. The scaling process is required if reduced system preconditioning is used.

xmdcheck

Check the matrix structure. This subroutine checks assigned array sizes and formats of *ia,ja* data structure.

Sequence of Calling Subroutines

Figure C1 illustrates a flow chart to use χMD package. In order to minimize computational work, the numerical factorization modules (xmdnfctr) should be used after obtaining solutions if the new coefficient matrix is assumed close to the coefficient matrix that is used at the previous calling of the symbolic preconditioner, such as xmdprecl. This case could happen when solving a transient problem. With a constant time step, the coefficient matrix does not change and symbolic and numeric factorizations of the coefficient matrix need only be performed once, as long as the user does not change entries in the factorized matrix. However, certain entries in the coefficient matrix will change with a change in the size of the time step. In the case of variable time step sizes, if the size does not change by an order of magnitude, the user can assume that the assembled coefficient matrix is close to the previous coefficient matrix, and consequently can call the numerical factorization modules directly. In this case in order to update the symbolic factorization, the symbolic factorization should be performed at least every fifth to eighth time step, or whenever the current time step is one order of magnitude greater than the last time step at which the symbolic factorization was performed.

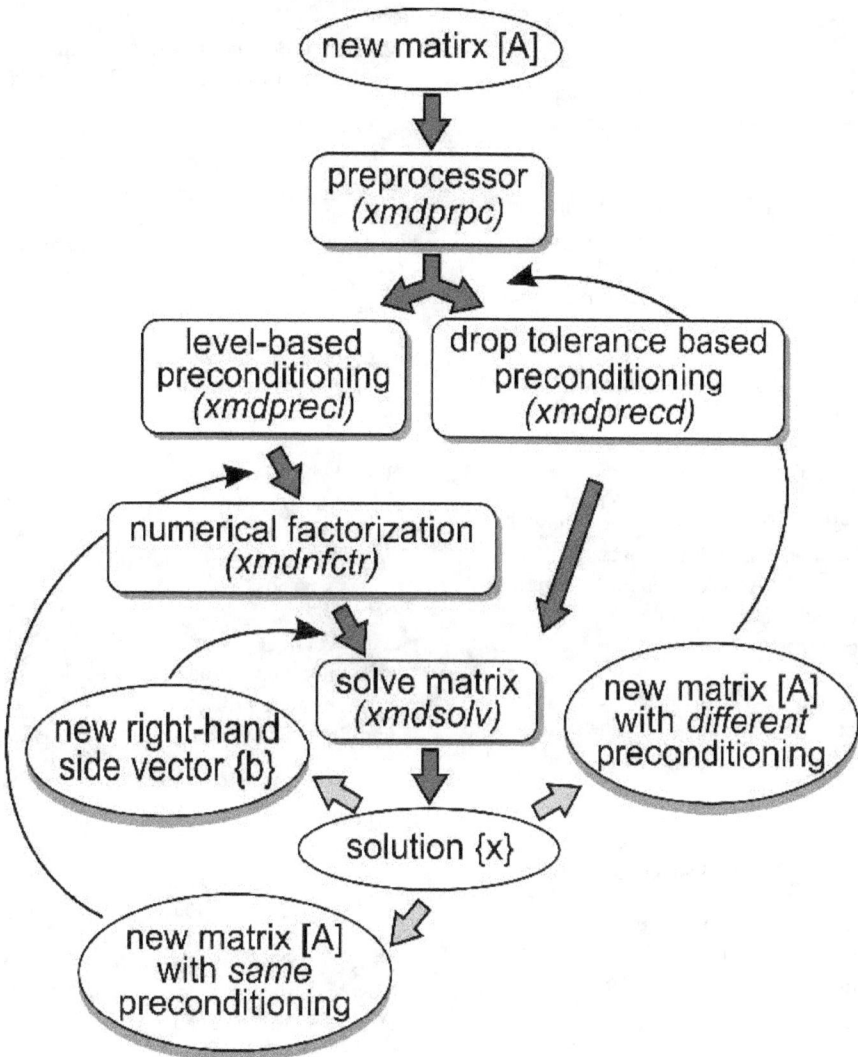

Figure C1.   Sequence of calling subroutines in χMD.

Selected References

Axelsson, O., 1977, Solution of linear systems of equations—Iterative methods: Berlin, Heidelberg, New York, Springer-Verlag.

Barrett, R., Berry, M., Chan, T.F., Demmel, J., Donato, J., Dongarra, J., Eijkhout, V., Pozo, R., Romine, C., and van der Vorst, H., 1994, Templates for the solution of linear systems—Building blocks for iterative methods: Philadelphia, SIAM.

Behie, G.A., and Forsyth, P.A., 1984, Incomplete factorization methods for fully implicit simulation of enhanced oil recovery: SIAM Journal on Scientic and Statistical Computing, v. 5, no. 3, p. 543-561.

Cuthill, E., and McKee, J., 1969, Reducing the bandwidth of sparse symmetric matrices, *in* Proceedings of the 24th National Conference of the Associations for Computing Machinery, New Jersey: Associations for Computing Machinery, Brandon Press, p. 157-172.

Eisenstat, S.C., Gursky, M.C., Schultz, M.H., and Sherman, A.H., 1982, Yale sparse matrix package i: The symmetric codes: International Journal for Numerical Methods in Engineering, no. 18, p. 1145-1151.

Duff, I.S., Erisman, A., and Reid, J., 1986, Direct methods for sparse matrices: Oxford, Oxford Science Publications.

George, A., 1971, Computer implementation of the nite-element method: Stanford, California, Stanford University, Department of Computer Sciences, Report STAN Cs-71-208.

Hestenes, M.R., and Stiefel, E., 1952, Methods of conjugate gradients for solving linear systems: Journal of Research of the National Bureau of Standards, no. 49, p. 409-436.

Munksgaard, N., 1980, Solving sparse symmetric sets of linear equations by preconditioned conjugate gradients: ACM Transactions on Mathematical Software, no. 6, p. 206-219.

Simon, H.D., 1988, Incomplete LU preconditioners for conjugate gradient-type iterative methods: SPE Reservoir Engineering, no. 3, p. 302-6.

Tinney, W.F., and Walker, J.W., 1967, Direct solutions of sparse network equations by optimally ordered triangular factorization: Proceedings, v. 55, p. 1801-1809.

Young, D.P., Melvin, R.G., Johnson, F.T., Bussoletti, J.E., Wigton, L.B., and Samant, S.S., 1989, 10: Application of sparse matrix solvers as effective preconditioners: SIAM Journal on Scientific and Statistical Computing, no. 6, p. 1186-1199.

van der Vorst, H.A., 1992, March, Bi-CGSTAB: A fast and smoothly converging variant of Bi-CG for the solution of nonsymmetric linear systems: SIAM Journal on Scientific and Statistical Computing, v. 13, no. 2, p. 631-644.

Vinsome, P.K.W., 1976, ORTHOMIN—An iterative method for solving sparse banded set of simultaneous linear equations, *in* the Fourth SPE Symposium on Numerical Simulation of Reservoir Performance, Los Angeles, Feb. 19-20, p. 149-159, SPE. paper SPE 5729.

Zlatev, Z., 1982, Use of iterative refinement in the solution of sparse linear systems: SIAM Journal on Numerical Analysis, no. 19, p. 381-399.

www.ingramcontent.com/pod-product-compliance
Lightning Source LLC
Chambersburg PA
CBHW081620170526
45166CB00009B/3045